Sparse Estimation with Math and R

Joe Suzuki

Sparse Estimation with Math and R

100 Exercises for Building Logic

Springer

Joe Suzuki
Graduate School of Engineering Science
Osaka University
Toyonaka, Osaka, Japan

ISBN 978-981-16-1445-3   ISBN 978-981-16-1446-0 (eBook)
https://doi.org/10.1007/978-981-16-1446-0

This Springer imprint is published by the registered company Springer Nature Singapore Pte Ltd.
The registered company address is: 152 Beach Road, #21-01/04 Gateway East, Singapore 189721,
Singapore

Preface

I started considering the sparse estimation problems around 2017 when I moved from the mathematics department to statistics in Osaka University, Japan. I have been studying information theory and graphical models for over 30 years.

The first book I found is "Statistical Learning with Sparsity" by T. Hastie, R. Tibshirani, and M. Wainwright. I thought it was a monograph rather than a textbook and that it would be tough for a nonexpert to read it through. However, I downloaded more than 50 papers that were cited in the book and read them all. In fact, the book does not instruct anything but only suggests how to study sparsity. The contrast between statistics and convex optimization is gradually attracting me as I understand the material.

On the other hand, it seems that the core results on sparsity have come out around 2010–2015 for research. However, I still think further possibilities and expansions are there. This book contains all the mathematical derivations and source programs, so graduate students can construct any procedure from scratch by getting help from this book.

Recently, I published books "Statistical Learning with Math and R" (SLMR) and "Statistical Learning with Math and Python" (SLMP) and will publish "Sparse Estimation with Math and Python" (SEMP). The common idea is behind the books (XXMR/XXMP). They not only give knowledge on statistical learning and sparse estimation but also help build logic in your brain by following each step of the derivations and each line of the source programs. I often meet with data scientists engaged in machine learning and statistical analyses for research collaborations and introduce my students to them. I recently found out that almost all of them think that (mathematical) logic rather than knowledge and experience is the most crucial ability for grasping the essence in their jobs. Our necessary knowledge is changing every day and can be obtained when needed. However, logic allows us to examine whether each item on the Internet is correct and follow any changes; we might miss even chances without it.

What Makes SEMR Unique?

I have summarized the features of this book as follows:

1. Developing logic
   To grasp the essence of the subject, we mathematically formulate and solve each ML problem and build those programs. The SEMR instills "logic" in the minds of the readers. The reader will acquire both the knowledge and ideas of ML, so that even if new technology emerges, they will be able to follow the changes smoothly. After solving the 100 problems, most of the students would say "I learned a lot".

2. Not just a story
   If programming codes are available, you can immediately take action. It is unfortunate when an ML book does not offer the source codes. Even if a package is available, if we cannot see the inner workings of the programs, all we can do is input data into those programs. In SEMR, the program codes are available for most of the procedures. In cases where the reader does not understand the math, the codes will help them understand what it means.

3. Not just a how to book: an academic book written by a university professor.
   This book explains how to use the package and provides examples of executions for those who are not familiar with them. Still, because only the inputs and outputs are visible, we can only see the procedure as a black box. In this sense, the reader will have limited satisfaction because they will not be able to obtain the essence of the subject. SEMR intends to show the reader the heart of ML and is more of a full-fledged academic book.

4. Solve 100 exercises: problems are improved with feedback from university students
   The exercises in this book have been used in university lectures and have been refined based on feedback from students. The best 100 problems were selected. Each chapter (except the exercises) explains the solutions, and you can solve all of the exercises by reading the book.

5. Self-contained
   All of us have been discouraged by phrases such as "for the details, please refer to the literature XX". Unless you are an enthusiastic reader or researcher, nobody will seek out those references. In this book, we have presented the material in such a way that consulting external references is not required. Additionally, the proofs are simple derivations, and the complicated proofs are given in the appendices at the end of each chapter. SEMR completes all discussions, including the appendices.

6. Readers' pages: questions, discussion, and program files. The reader can ask any question on the book via https://bayesnet.org/books.

Osaka, Japan   Joe Suzuki
March 2021

Acknowledgments The author wishes to thank Tianle Yang, Ryosuke Shinmura, and Tomohiro Kamei for checking the manuscript in Japanese. This English book is largely based on the Japanese book published by Kyoritsu Shuppan Co., Ltd. in 2020. The author would like to thank Kyoritsu Shuppan Co., Ltd., in particular, its editorial members Mr. Tetsuya Ishii and Ms. Saki Otani. The author also appreciates Ms. Mio Sugino, Springer, for preparing the publication and providing advice on the manuscript.

Contents

Chapter 1
Linear Regression

In general statistics, we often assume that the number of samples N is greater than the number of variables p. If this is not the case, it may not be possible to solve for the best-fitting regression coefficients using the least squares method, or it is too computationally costly to compare a total of 2^p models using some information criterion.

When p is greater than N (also known as the sparse situation), even for linear regression, it is more common to minimize, instead of the usual squared error, the modified objective function to which a term is added to prevent the coefficients from being too large (the so-called regularization term). If the regularization term is a constant λ times the L1-norm (resp. L2-norm) of the coefficient vector, it is called Lasso (resp. Ridge). In the case of Lasso, if the value of λ increases, there will be more coefficients that go to 0, and when λ reaches a certain value, all the coefficients will eventually become 0. In that sense, we can say that Lasso also plays a role in model selection.

In this chapter, we examine the properties of Lasso in comparison to those of Ridge. After that, we investigate the elastic net, a regularized regression method that combines the advantages of both Ridge and Lasso. Finally, we consider how to select an appropriate value of λ.

1.1 Linear Regression

Throughout this chapter, let $N \geq 1$ and $p \geq 1$ be integers, and let the (i, j) element of the matrix $X \in \mathbb{R}^{N \times p}$ and the k-th element of the vector $y \in \mathbb{R}^N$ be denoted by $x_{i,j}$ and y_k, respectively. Using these X, y, we find the intercept $\beta_0 \in \mathbb{R}$ and the slope $\beta = [\beta_1, \ldots, \beta_p]^T$ that minimize $\|y - \beta_0 - X\beta\|^2$. Here, the L2-norm of $z = [z_1, \ldots, z_N]^T$ is denoted by $\|z\| := \sqrt{\sum_{i=1}^{N} z_i^2}$.

© The Author(s), under exclusive license to Springer Nature Singapore Pte Ltd. 2021
J. Suzuki, *Sparse Estimation with Math and R*,
https://doi.org/10.1007/978-981-16-1446-0_1

First, for the sake of simplicity, we assume that the j-th column of X, ($j = 1, \ldots, p$), and y have already been centered. That is, for each $j = 1, \ldots, p$, define $\bar{x}_j := \frac{1}{N} \sum_{i=1}^{N} x_{i,j}$, and assume that $\bar{x}_j$ has already been subtracted from each $x_{i,j}$ so that $\bar{x}_j = 0$ is satisfied. Similarly, defining $\bar{y} := \frac{1}{N} \sum_{i=1}^{N} y_i$, we assume that $\bar{y}$ was subtracted in advance from each y_i so that $\bar{y} = 0$ holds. Under this condition, one of the parameters $(\hat{\beta}_0, \hat{\beta})$ for which we need to solve, say, $\hat{\beta}_0$, is always 0. In particular,

$$0 = \frac{\partial}{\partial \beta_0} \sum_{i=1}^{N} (y_i - \beta_0 - \sum_{j=1}^{p} x_{i,j} \beta_j)^2 = -2N(\bar{y} - \beta_0 - \sum_{j=1}^{p} \bar{x}_j \beta_j) = 2N\beta_0$$

holds. Thus, from now, without loss of generality, we may assume that the intercept β_0 is zero and use this in our further calculations.

We begin by first observing the following equality:

$$\begin{bmatrix} \dfrac{\partial}{\partial \beta_1} \sum_{i=1}^{N} (y_i - \sum_{k=1}^{p} \beta_k x_{i,k})^2 \\ \vdots \\ \dfrac{\partial}{\partial \beta_p} \sum_{i=1}^{N} (y_i - \sum_{k=1}^{p} \beta_k x_{i,k})^2 \end{bmatrix} = -2 \begin{bmatrix} x_{1,1} & \cdots & x_{N,1} \\ vdots & \ddots & \vdots \\ x_{1,p} & \cdots & x_{N,p} \end{bmatrix} \begin{bmatrix} y_1 - \sum_{k=1}^{p} \beta_k x_{1,k} \\ \vdots \\ y_N - \sum_{k=1}^{p} \beta_k x_{N,k} \end{bmatrix}$$

$$= -2X^T (y - X\beta). \tag{1.1}$$

In particular, the j-th element of each side can be rewritten as

$$-2 \sum_{i=1}^{N} x_{i,j} (y_i - \sum_{k=1}^{p} \beta_k x_{i,k}).$$

Thus, when we set the right-hand side of (1.1) equal to 0, and if $X^T X$ is invertible, then β becomes

$$\hat{\beta} = (X^T X)^{-1} X^T y. \tag{1.2}$$

For the case where $p = 1$, write each column of X as $x_1, \ldots, x_N$, we see that

$$\hat{\beta} = \frac{\sum_{i=1}^{N} x_i y_i}{\sum_{i=1}^{N} x_i^2}. \tag{1.3}$$

If we had not performed the data centering, we would still obtain the same slope $\hat{\beta}$, though the intercept $\hat{\beta}_0$ would be

$$\hat{\beta}_0 = \bar{y} - \sum_{j=1}^{p} \bar{x}_j \hat{\beta}_j. \tag{1.4}$$

Here, $\bar{x}_j$ ($j = 1, \ldots, p$) and $\bar{y}$ are the means before data centering.

We can implement the above using the R language as follows:

```
1  inner.prod = function(x, y) return(sum(x * y))
2  linear = function(X, y) {
3    n = nrow(X); p = ncol(X)
4    X = as.matrix(X); x.bar = array(dim = p); for (j in 1:p) x.bar[j] = mean(X
       [, j])
5    for (j in 1:p) X[, j] = X[, j] - x.bar[j]        ## mean-centering for X
6    y = as.vector(y); y.bar = mean(y); y = y - y.bar   ## mean-centering for y
7    beta = as.vector(solve(t(X) %*% X) %*% t(X) %*% y)
8    beta.0 = y.bar - sum(x.bar * beta)
9    return(list(beta = beta, beta.0 = beta.0))
10 }
```

In this book, we focus more on the sparse case, i.e., when p is larger than N. In this case, a problem arises. When $N < p$, the matrix $X^T X$ does not have an inverse. In fact, since

$$\text{rank}(X^T X) \le \text{rank}(X) \le \min\{N, p\} = N < p,$$

$X^T X$ is singular. Moreover, when X has precisely the same two columns, $\text{rank}(X) < p$, and the inverse matrix does not exist.

On the other hand, if p is rather large, we have p independent variables to choose from for the predictors of the variable when carrying out model selection. Thus, the combinations are

$$\{\}, \{1\}, \{2\}, \ldots, \{1, \ldots, p\}$$

(that is, whether we choose each of the variables or not). This means we have to compare a total of 2^p models. Then, to extract the proper model combination by using an information criterion or cross-validation, the computational resources required will grow exponentially with the number of variables p.

To deal with this kind of problem, let us consider the following. Let $\lambda \ge 0$ be a constant. We add a term to $\|y - X\beta\|^2$ that penalizes β for being too large in size. Specifically, we define

$$L := \frac{1}{2N}\|y - X\beta\|^2 + \lambda\|\beta\|_1 \tag{1.5}$$

or

$$L := \frac{1}{N}\|y - X\beta\|^2 + \lambda\|\beta\|_2^2. \tag{1.6}$$

Our work now is to solve for the minimizer $\beta \in \mathbb{R}^p$ to one of the above quantities. Here, for $\beta = [\beta_1, \ldots, \beta_p]^T$, $\|\beta\|_1 := \sum_{j=1}^{p} |\beta_j|$ is the L1-norm and $\|\beta\|_2 := \sqrt{\sum_{j=1}^{p} \beta_j^2}$ is the L2-norm of β. To be more specific, we plug mean-centered

$X \in \mathbb{R}^{N \times p}$ and $y \in \mathbb{R}^N$ into either (1.5) or (1.6), then minimize it with respect to the slope $\hat{\beta} \in \mathbb{R}^p$ (this is called Lasso or Ridge, respectively), and finally we use (1.4) to compute $\hat{\beta}_0$.

1.2 Subderivative

To address the minimization problem for Lasso, we need a method for optimizing functions that are not differentiable. When we want to find the points x of the maxima or minima of a single-variable polynomial, say, $f(x) = x^3 - 2x + 1$, we can differentiate it and find the solution to $f'(x) = 0$. However, what should we do when we encounter functions such as $f(x) = x^2 + x + 2|x|$, which contain an absolute value? To address this, we need to extend our concept of differentiation to a more general one.

Throughout the following claims, let us assume that f is convex [4, 6]. In general, we say that f is convex (downward)[1] if, for any $0 < \alpha < 1$ and $x, y \in \mathbb{R}$,

$$f(\alpha x + (1 - \alpha)y) \leq \alpha f(x) + (1 - \alpha)f(y)$$

holds. For instance, $f(x) = |x|$ is convex (Fig. 1.1, left) because

$$|\alpha x + (1 - \alpha)y| \leq \alpha|x| + (1 - \alpha)|y|$$

is satisfied. To check this, since both sides are nonnegative, the RHS squared minus the LHS squared gives $2\alpha(1 - \alpha)(|xy| - xy) \geq 0$. As another example, consider

$$f(x) = \begin{cases} 1, & x \neq 0 \\ 0, & x = 0. \end{cases} \tag{1.7}$$

This function satisfies the following:

$$f(\alpha \cdot 0 + (1 - \alpha) \cdot 1) = 1 > 1 - \alpha = \alpha f(0) + (1 - \alpha)f(1)$$

Therefore, it is not convex (Fig. 1.1, right). If functions f, g are convex, then for any $\beta, \gamma \geq 0$, the function $\beta f(x) + \gamma g(x)$ has to be convex since the following holds:

$$\begin{aligned} \beta\{f(\alpha x + (1 - \alpha)y)\} + \gamma\{g(\alpha x + (1 - \alpha)y)\} \\ \leq \alpha\beta f(x) + (1 - \alpha)\beta f(y) + \alpha\gamma g(x) + (1 - \alpha)\gamma g(y) \\ = \alpha\{\beta f(x) + \gamma g(x)\} + (1 - \alpha)\{\beta f(y) + \gamma g(y)\}. \end{aligned}$$

---

[1] Throughout this book, when we refer to convex, we mean convex downward.

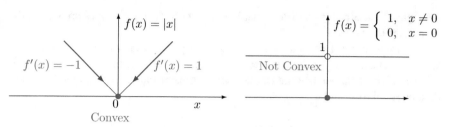

Fig. 1.1 Left: $f(x) = |x|$ is convex. However, at the origin, the derivatives from each side differ; thus, it is not differentiable. Right: We cannot simply judge from its shape, but this function is not convex

Next, for any convex function $f : \mathbb{R} \to \mathbb{R}$, fix $x_0 \in \mathbb{R}$ arbitrarily. For all $x \in \mathbb{R}$, we say that the set of all $z \in \mathbb{R}$ that satisfies

$$f(x) \geq f(x_0) + z(x - x_0) \tag{1.8}$$

is a subderivative of f at x_0.

If f is differentiable at x_0, then the subderivative will be a set that contains only one element, say, $f'(x_0)$.[2] We prove this as follows.

First, the convex function f is differentiable at x_0; thus, it satisfies $f(x) \geq f(x_0) + f'(x_0)(x - x_0)$. To see this, since f is convex,

$$f(\alpha x + (1 - \alpha)x_0) \leq \alpha f(x) + (1 - \alpha)f(x_0).$$

This can be rewritten as

$$f(x) \geq f(x_0) + \frac{f(x_0 + \alpha(x - x_0)) - f(x_0)}{\alpha(x - x_0)}(x - x_0).$$

In fact, whether $x < x_0$ or $x > x_0$, we have that

$$\lim_{\alpha \searrow 0} \frac{f(x_0 + \alpha(x - x_0)) - f(x_0)}{\alpha(x - x_0)} = f'(x_0)$$

holds. Thus, the above in equation is true.

Next, when the convex function f is differentiable at x_0, we can show that $f'(x_0)$ is the one and only value of z that satisfies (1.8). In particular, when $x > x_0$, for (1.8) to be satisfied, we need $\dfrac{f(x) - f(x_0)}{x - x_0} \geq z$. Similarly, when $x < x_0$, for (1.8) to be satisfied, we need $\dfrac{f(x) - f(x_0)}{x - x_0} \leq z$. Thus, z needs to be larger than or equal to the derivative on the left and, at the same time, be less than or equal to the derivative on

---

[2] In this case, we would not write it as a set $\{f'(x_0)\}$ but as an element $f'(x_0)$.

the right at x_0. Since f is differentiable at x_0, those two derivatives are equal; this completes the proof.

The main interest of this book is specifically the case where $f(x) = |x|$ and $x_0 = 0$. Hence, by (1.8), its subderivative is the set of z such that for any $x \in \mathbb{R}$, $|x| \geq zx$. These values of z lie in the interval greater than or equal to -1 and less than or equal to 1, and

$$\text{for any arbitrary } x, \quad |x| \geq zx \iff |z| \leq 1$$

is true. Let us confirm this. If for any x, $|x| \geq zx$ holds, then for $x > 0$ and $x < 0$, $z \leq 1$ and $z \geq -1$, respectively, need to be true. Conversely, if $-1 \leq z \leq 1$, then $zx \leq |z||x| \leq |x|$ is true for any arbitrary $x \in \mathbb{R}$.

Example 1 By dividing into three cases $x < 0$, $x = 0$, and $x > 0$, find the values x that attain the minimum of $x^2 - 3x + |x|$ and $x^2 + x + 2|x|$. Note that for $x \neq 0$, we can find their usual derivatives, but for $f(x) = |x|$ at $x = 0$, its subderivative is the interval $[-1, 1]$.

$$x^2 - 3x + |x| = \begin{cases} x^2 - 3x + x, & x \geq 0 \\ x^2 - 3x - x, & x < 0 \end{cases} = \begin{cases} x^2 - 2x, & x \geq 0 \\ x^2 - 4x, & x < 0 \end{cases}$$

$$(x^2 - 3x + |x|)' = \begin{cases} 2x - 2, & x > 0 \\ 2x - 3 + [-1, 1] = -3 + [-1, 1] = [-4, -2] \not\ni 0, & x = 0 \\ 2x - 4 < 0, & x < 0. \end{cases}$$

Therefore, $x^2 - 3x + |x|$ has a minimum at $x = 1$ (Fig. 1.2, left).

$$x^2 + x + 2|x| = \begin{cases} x^2 + x + 2x, & x \geq 0 \\ x^2 + x - 2x, & x < 0 \end{cases} = \begin{cases} x^2 + 3x, & x \geq 0 \\ x^2 - x, & x < 0 \end{cases}$$

$$(x^2 + x + 2|x|)' = \begin{cases} 2x + 3 > 0, & x > 0 \\ 2x + 1 + 2[-1, 1] = 1 + 2[-1, 1] = [-1, 3] \ni 0, & x = 0 \\ 2x - 1 < 0, & x < 0. \end{cases}$$

Therefore, $x^2 + x + 2|x|$ has a minimum at $x = 0$ (Fig. 1.2, right). We use the following code to draw the figures.

```
1  curve(x ^ 2 - 3 * x + abs(x), -2, 2, main = "y = x^2 - 3x + |x|")
2  points(1, -1, col = "red", pch = 16)
3  curve(x ^ 2 + x + 2 * abs(x), -2, 2, main = "y = x^2 + x + 2|x|")
4  points(0, 0, col = "red", pch = 16)
```

The subderivative of $f(x) = |x|$ at $x = 0$ is the interval $[-1, 1]$. This fact summarizes this chapter.

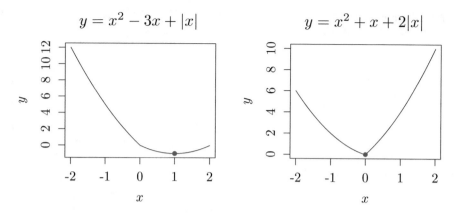

Fig. 1.2 $x^2 - 3x + |x|$ (left) has a minimum at $x = 1$, and $x^2 + x + 2|x|$ (right) has a minimum at $x = 0$. Neither is differentiable at $x = 0$. Despite not being differentiable, the point is a minimum for the figure on the right

1.3 Lasso

As stated in Sect. 1.1, the method considered for the minimization of

$$L := \frac{1}{2N} \|y - X\beta\|^2 + \lambda\|\beta\|_1 \tag{1.5}$$

is called Lasso [28].

From the formularization of (1.5) and (1.6), we can tell that Lasso and Ridge are the same in the sense that they both try to control the size of regression coefficients β. However, Lasso also has the property of leaving significant coefficients as non-negative, which is particularly beneficial in variable selection. Let us consider its mechanism.

Note that in (1.5), the division of the first term by 2 is not essential: we would obtain an equivalent formularization if we double the value of λ. For the sake of simplicity, first let us assume that

$$\frac{1}{N}\sum_{i=1}^{N} x_{i,j}x_{i,k} = \begin{cases} 1, & j = k \\ 0, & j \neq k \end{cases} \tag{1.9}$$

holds, and let $s_j = \frac{1}{N}\sum_{i=1}^{N} x_{i,j}y_i$. With this assumption, the calculations are made much simpler.

Fig. 1.3 Shape of the
function $\mathcal{S}_\lambda(x)$ when $\lambda = 5$

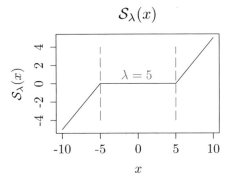

Solving for the subderivative of L with respect to β_j gives

$$0 \in -\frac{1}{N}\sum_{i=1}^{N} x_{i,j}\left(y_i - \sum_{k=1}^{p} x_{i,k}\beta_k\right) + \lambda \begin{cases} 1, & \beta_j > 0 \\ [-1,1], & \beta_j = 0 \\ -1, & \beta_j < 0, \end{cases} \qquad (1.10)$$

which means that

$$0 \in \begin{cases} -s_j + \beta_j + \lambda, & \beta_j > 0 \\ -s_j + \beta_j + \lambda[-1,1], & \beta_j = 0 \\ -s_j + \beta_j - \lambda, & \beta_j < 0. \end{cases}$$

Thus, we have that

$$\beta_j = \begin{cases} s_j - \lambda, & s_j > \lambda \\ 0, & -\lambda \le s_j \le \lambda \\ s_j + \lambda, & s_j < -\lambda. \end{cases}$$

Here, the RHS can be rewritten using the following function:

$$\mathcal{S}_\lambda(x) = \begin{cases} x - \lambda, & x > \lambda \\ 0, & -\lambda \le x \le \lambda \\ x + \lambda, & x < -\lambda, \end{cases} \qquad (1.11)$$

and hence becomes $\beta_j = \mathcal{S}_\lambda(s_j)$. We plot the graph of $\mathcal{S}_\lambda(x)$ when $\lambda = 5$ in Fig. 1.3 using the code provided below:

```
soft.th = function(lambda, x) return(sign(x) * pmax(abs(x) - lambda, 0))
curve(soft.th(5, x), -10, 10, main = "soft.th(lambda, x)")
segments(-5, -4, -5, 4, lty = 5, col = "blue"); segments(5, -4, 5, 4, lty =
    5, col = "blue")
text(-0.2, 1, "lambda = 5", cex = 1.5, col = "red")
```

Next, let us consider the case where (1.9) is not satisfied. We rewrite (1.10) by

$$0 \in -\frac{1}{N} \sum_{i=1}^{N} x_{i,j}(r_{i,j} - x_{i,j}\beta_j) + \lambda \begin{cases} 1, & \beta_j > 0 \\ [-1, 1], & \beta_j = 0 \\ -1, & \beta_j < 0. \end{cases}$$

Here, we denote $y_i - \sum_{k \neq j} x_{i,k}\beta_k$ by $r_{i,j}$, and let $\frac{1}{N} \sum_{i=1}^{N} r_{i,j}x_{i,j}$ be s_j. Next, fix β_k $(k \neq j)$, and update β_j. We do this repeatedly from $j = 1$ to $j = p$ until it converges (coordinate descent). For example, we can implement the algorithm as follows:

```
1  linear.lasso = function(X, y, lambda = 0, beta = rep(0, ncol(X))) {
2    n = nrow(X); p = ncol(X)
3    res = centralize(X, y)    ## centering (please refer to below code)
4    X = res$X; y = res$y
5    eps = 1; beta.old = beta
6    while (eps > 0.001) {    ## wait for this loop to converge
7      for (j in 1:p) {
8        r = y - as.matrix(X[, -j]) %*% beta[-j]
9        beta[j] = soft.th(lambda, sum(r * X[, j]) / n) / (sum(X[, j] * X[, j])
           / n)
10     }
11     eps = max(abs(beta - beta.old)); beta.old = beta
12   }
13   beta = beta / res$X.sd    ## restore each variable coefficient to that of
         before standardized
14   beta.0 = res$y.bar - sum(res$X.bar * beta)
15   return(list(beta = beta, beta.0 = beta.0))
16 }
```

Note that here, after we obtain the value of β, we use $\bar{x}_j$ $(j = 1, \ldots, p)$ and $\bar{y}$ to calculate the value of β_0. The following centralize function performs data centering and returns a list of five results.

```
1  centralize = function(X, y, standardize = TRUE) {
2    X = as.matrix(X)
3    n = nrow(X); p = ncol(X)
4    X.bar = array(dim = p)              ## mean of each comlumn of X
5    X.sd = array(dim = p)               ## standard deviation of each column of X
6    for (j in 1:p) {
7      X.bar[j] = mean(X[, j])
8      X[, j] = (X[, j] - X.bar[j])      ## centering each column of X
9      X.sd[j] = sqet(var(X[, j]))
10     if (standardize == TRUE) X[, j] = X[, j] / X.sd[j]    ## standardize each
         column of X
11   }
12   if (is.matrix(y)) {              ## the case when y is a matrix
13     K = ncol(y)
14     y.bar = array(dim = K)           ## mean of y
15     for (k in 1:K) {
16       y.bar[k] = mean(y[, k])
17       y[, k] = y[, k] - y.bar[k]    ## centering y
18     }
```

```
19  } else {                          ## the case when y is a vector
20    y.bar = mean(y)
21    y = y - y.bar
22  }
23  return(list(X = X, y = y, X.bar = X.bar, X.sd = X.sd, y.bar = y.bar))
24 }
```

Thus, we may standardize the data first, then perform Lasso, and finally restore the data. The aim of doing this is to examine our data all at once. Because the algorithm sets all $\hat{\beta}_j$ less than or equal to λ to 0, we do not want it to be based on the indices $j = 1, \ldots, p$. Each j-th column of X is divided by scale[j], and consequently the estimated β_j will be larger to that extent. We then divide β_j by scale[j] as well.

Example 2 Putting U.S. crime data https://web.stanford.edu/~hastie/StatLearn Sparsity/data.html into the text file crime.txt, we set the crime rate per 1 million residents as the target variable and then select appropriate explanatory variables from the list below by performing Lasso.

Column	Cov./Res.	Definition of Variable
1	response	crime rate per 1 million residents
2		(we currently do not use this)
3	covariate	annual police funding
4	covariate	% of people 25 years+ with 4 yrs. of high school education
5	covariate	% of 16–19-year-old persons not in high school and not high school graduates
6	covariate	% of 18–24-year-old persons in college
7	covariate	% of people 25 years+ with at least 4 years of college education

We call the function linear.lasso and execute as described below:

```
1  df = read.table("crime.txt")
2  x = df[, 3:7]; y = df[, 1]; p = ncol(x); lambda.seq = seq(0, 200, 0.1)
3  plot(lambda.seq, xlim = c(0, 200), ylim = c(-10, 20), xlab = "lambda", ylab =
       "beta",
4      main = "each coefficient's value for each lambda", type = "n", col = "
       red")
5  for (j in 1:p) {
6    coef.seq = NULL
7    for (lambda in lambda.seq) coef.seq = c(coef.seq, linear.lasso(x, y, lambda
       )$beta[j])
8    par(new = TRUE); lines(lambda.seq, coef.seq, col = j)
9  }
10 legend("topright",
11         legend = c("annual police funding", "\% of people 25 years+ with 4 yrs
       . of high school",
12                     "\% of 16--19 year-olds not in highschool and not
       highschool graduates",
13                     "\% of people 25 years+ with at least 4 years of college"),
14       col = 1:p, lwd = 2, cex = .8)
```

The Values of Coefficients for each λ

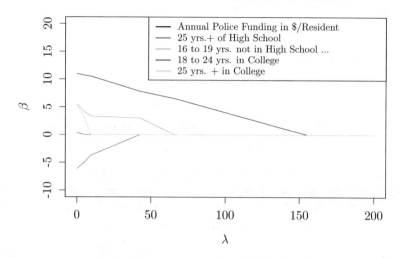

Fig. 1.4 Result of Example 2. In the case of Lasso, we see that as λ increases, the coefficients decrease. At a certain λ, all the coefficients will be 0. The λ at which each coefficient becomes 0 varies

As we can see in Fig. 1.4, as λ increases, the absolute value of each coefficient decreases. When λ reaches a certain value, all coefficients go to 0. In other words, for each value of λ, the set of nonzero coefficients differs. The larger λ becomes, the smaller the set of selected variables.

When performing coordinate descent, it is quite common to begin with a λ large enough that every coefficient is zero and then make λ smaller gradually. This method is called a warm start, which utilizes the fact that when we want to calculate the coefficient β for all values of λ, we can improve the calculation performance by setting the initial value of β to the estimated β from a previous λ. For example, we can write the program as follows:

```
1  warm.start = function(X, y, lambda.max = 100) {
2    dec = round(lambda.max / 50); lambda.seq = seq(lambda.max, 1, -dec)
3    r = length(lambda.seq); p = ncol(X); coef.seq = matrix(nrow = r, ncol = p)
4    coef.seq[1, ] = linear.lasso(X, y, lambda.seq[1])$beta
5    for (k in 2:r) coef.seq[k, ] = linear.lasso(X, y, lambda.seq[k], coef.seq[(
       k - 1), ])$beta
6    return(coef.seq)
7  }
```

Example 3 We use the warm start method to reproduce the coefficient β for each λ in Example 2.

```
1 crime = read.table("crime.txt"); X = crime[, 3:7]; y = crime[, 1]
2 coef.seq = warm.start(X, y, 200)
3 p = ncol(X); lambda.max = 200; dec = round(lambda.max / 50)
4 lambda.seq = seq(lambda.max, 1, -dec)
5 plot(log(lambda.seq), coef.seq[, 1], xlab = "log(lambda)", ylab = "
    coefficients",
6     ylim = c(min(coef.seq), max(coef.seq)), type = "n")
7 for (j in 1:p) lines(log(lambda.seq), coef.seq[, j], col = j)
```

First, we make the value of λ large enough so that all β_j $(j = 1, \ldots, p)$ are 0 and then gradually decrease the size of λ while performing coordinate descent. Here, for simplicity, we assume that for each $j = 1, \ldots, p$, we have $\sum_{i=1}^{N} x_{i,j}^2 = 1$; moreover, the values of $\sum_{i=1}^{N} x_{i,j} y_i$ are all different. In this case, the smallest λ that makes $\beta_j = 0$ $(j = 1, \ldots, p)$ can be calculated by $\lambda = \max_{1 \le j \le p} \left| \frac{1}{N} \sum_{i=1}^{N} x_{i,j} y_i \right|$. Particularly, for any λ larger than this formula, it will be satisfied that for all $j = 1, \ldots, p, \beta_j = 0$. Then, we have that $r_{i,j} = y_i$ $(i = 1, \ldots, N)$ and

$$-\lambda \le -\frac{1}{N} \sum_{i=1}^{N} x_{i,j}(r_{i,j} - x_{i,j}\beta_j) \le \lambda$$

hold. Thus, when we decrease the size of λ, one of the values of j will satisfy $\frac{1}{N}|\sum_{i=1}^{N} x_{i,j}(r_{i,j} - x_{i,j}\beta_j)| = \lambda$. Again, since $\beta_k = 0$ $(k \ne j)$, if we continue to make it smaller, we still have that $r_{i,j} = y_i$ $(i = 1, \ldots, N)$, and thus the value of $|\frac{1}{N} \sum_{i=1}^{N} x_{i,j} y_i|$ for that j becomes smaller than λ.

As a standard R package for Lasso, `glmnet` is often used [11].

Example 4 (Boston) Using the Boston dataset from the MASS package and setting the variable of the 14th column as the target variable and the other 13 variables as predictors, we plot a graph similar to that of the previous one (Fig. 1.5).

Fig. 1.5 The execution result of Example 4. The numbers at the top represent the number of estimated coefficients that are not zero

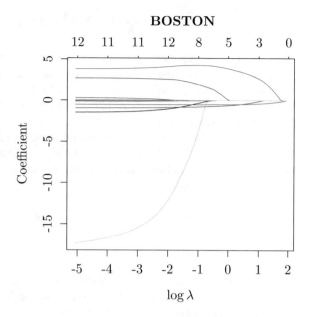

Column	Variable	Definition of Variable
1	CRIM	per capita crime rate by town
2	ZN	proportion of residential land zoned for lots over 25,000 sq.ft.
3	INDUS	proportion of nonretail business acres per town
4	CHAS	Charles River dummy variable (1 if tract bounds river; 0 otherwise)
5	NOX	nitric oxide concentration (parts per 10 million)
6	RM	average number of rooms per dwelling
7	AGE	proportion of owner-occupied units built prior to 1940
8	DIS	weighted distances to five Boston employment centers
9	RAD	index of accessibility to radial highways
10	TAX	full-value property-tax rate per $10,000
11	PTRATIO	pupil-teacher ratio by town
12	BLACK	proportion of blacks by town
13	LSTAT	% lower status of the population
14	MEDV	median value of owner-occupied homes in multiples of $1,000

```
1 library(glmnet); library(MASS)
2 df = Boston; x = as.matrix(df[, 1:13]); y = df[, 14]
3 fit = glmnet(x, y); plot(fit, xvar = "lambda", main = "BOSTON")
```

So far, we have seen how Lasso can be useful in the process of variable selection. However, we have not explained the reason why we seek to minimize (1.5) for $\beta \in \mathbb{R}^p$. Why do we not instead consider minimizing the following usual information criterion

$$\frac{1}{2N}\|y - X\beta\|^2 + \lambda\|\beta\|_0 \tag{1.12}$$

for β? Here, $\|\cdot\|_0$ represents the number of nonzero elements of that vector.

Lasso, as well as Ridge, which is to be discussed in the next chapter, has the advantage that minimization is convex. For a globally convex function, a local minimum is also considered the global minimum to find the optimal solution effectively. On the other hand, the minimization by (1.12) requires an exponential time for the number of variables, p. In particular, since the function in (1.7) is not convex, $\|\beta\|_0$ cannot be convex either. Because of this, we need to consider instead the minimization of (1.5). An optimization problem becomes meaningful only after there exists an effective search algorithm.

1.4 Ridge

In Sect. 1.1, we made an assumption about $X \in \mathbb{R}^{N \times p}$, $y \in \mathbb{R}^N$ that the matrix $X^T X$ is invertible, and, based on this, we showed that the β that minimizes the squared error $\|y - X\beta\|^2$ is given by $\hat{\beta} = (X^T X)^{-1} X^T y$.

First, when $N \geq p$, the possibility that $X^T X$ is singular is not that high, though we may have another problem instead: the confidence interval becomes significant when the determinant is small. To cope with this, we let $\lambda \geq 0$ be a constant and add to the squared error the norm of β times λ. That is, the method considering the minimization of

$$L := \frac{1}{N}\|y - X\beta\|^2 + \lambda\|\beta\|_2^2 \tag{1.6}$$

is commonly used. This method is called Ridge. Differentiating L with respect to β gives

$$0 = -\frac{2}{N} X^T (y - X\beta) + 2\lambda\beta.$$

If $X^T X + \lambda I$ is not singular, we obtain

$$\hat{\beta} = (X^T X + N\lambda I)^{-1} X^T y.$$

Here, whenever $\lambda > 0$, even for the case $N < p$, we have that $X^T X + N\lambda I$ is nonsingular. In particular, since the matrix $X^T X$ is positive semidefinite, we have that its eigenvalues $\mu_1, \ldots, \mu_p$ are all nonnegative. Therefore, the eigenvalues of $X^T X + N\lambda I$ can be calculated by

$$\det(X^T X + N\lambda I - tI) = 0 \implies t = \mu_1 + N\lambda, \ldots, \mu_p + N\lambda > 0,$$

and thus all of them are positive.

Again, when all the eigenvalues are positive, their product, $\det(X^T X + N\lambda I)$, is also positive, which is the same as saying that $X^T X + N\lambda I$ is nonsingular. Note that this always holds regardless of the sizes of p, N. When $N < p$, the rank of $X^T X \in \mathbb{R}^{p \times p}$ is less than or equal to N, and hence the matrix is singular. Therefore, for this case, the following conditions are equivalent:

$$\lambda > 0 \iff X^T X + N\lambda I \text{ is nonsingular.}$$

As an example of the Ridge case, we can write the following program:

```
1  ridge = function(X, y, lambda = 0) {
2    X = as.matrix(X); p = ncol(X); n = length(y)
3    res = centralize(X, y); X = res$X; y = res$y
4    ## the process of Ridge is only the next one line
5    beta = drop(solve(t(X) %*% X + n * lambda * diag(p)) %*% t(X) %*% y)
6    beta = beta / res$X.sd  ## restore data to those of before standardized
7    beta.0 = res$y.bar - sum(res$X.bar * beta)
8    return(list(beta = beta, beta.0 = beta.0))
9  }
```

Example 5 We use the same U.S. crime data as that of Example 2 and perform the following analysis. To control the size of the coefficient of each predictor, we call the function `ridge` and then execute.

```
1  df = read.table("crime.txt")
2  x = df[, 3:7]; y = df[, 1]; p = ncol(x); lambda.seq = seq(0, 100, 0.1)
3  plot(lambda.seq, xlim = c(0, 100), ylim = c(-10, 20), xlab = "lambda", ylab =
          "beta",
4        main = "each coefficient's value for each lambda", type = "n", col = "
          red")
5  for (j in 1:p) {
6    coef.seq = NULL
7    for (lambda in lambda.seq) coef.seq = c(coef.seq, ridge(X, y, lambda)$beta[
          j])
8    par(new = TRUE); lines(lambda.seq, coef.seq, col = j)
9  }
10 legend("topright",
11          legend = c("annual police funding", "\% of people 25 years+ with 4 yrs
          . of high school",
12                     "\% of 16--19 year-olds not in highschool and not
          highschool graduates",
13                     "\% of people 25 years+ with at least 4 years of college"),
14         col = 1:p, lwd = 2, cex = .8)
```

In Fig. 1.6, we plot how each coefficient changes with the value of λ.

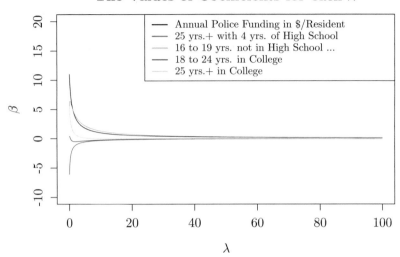

Fig. 1.6 The execution result of Example 5. The changes in the coefficient β with respect to λ based on Ridge. As λ becomes larger, each coefficient decreases to 0

```
crime = read.table("crime.txt")
X = crime[, 3:7]
y = crime[, 1]
linear(X, y)
```

```
$beta
[1] 10.9806703 -6.0885294  5.4803042  0.3770443  5.5004712
$beta.0
[1] 489.6486
```

```
ridge(X, y)
```

```
$beta
[1] 10.9806703 -6.0885294  5.4803042  0.3770443  5.5004712
$beta.0
[1] 489.6486
```

```
ridge(X, y, 200)
```

```
1   $beta
2   [1]  0.056351799 -0.019763974   0.077863094 -0.017121797 -0.007039304
3   $beta.0
4   [1] 716.4044
```

1.5 A Comparison Between Lasso and Ridge

Next, let us compare Fig. 1.4 of Lasso to Fig. 1.6 of Ridge. We can see that they are the same in the sense that when λ becomes larger, the absolute value of each coefficient approaches 0. However, in the case of Lasso, when λ reaches a certain value, one of the coefficients becomes exactly zero, and the time at which that occurs varies for each variable. Thus, Lasso can be used for variable selection.

So far, we have shown this fact analytically, but it is also good to have an intuition geometrically. Figures similar to those in Fig. 1.7 are widely used when one wants to compare Lasso and Ridge.

Let $p = 2$ such that $X \in \mathbb{R}^{N \times p}$ is composed of two columns $x_{i,1}, x_{i,2}$ ($i = 1, \ldots, N$). In the least squares process, we solve for the β_1, β_2 values that minimize $S := \sum_{i=1}^{N} (y_i - \beta_1 x_{i,1} - \beta_2 x_{i,2})^2$. For now, let us denote them by $\hat{\beta}_1, \hat{\beta}_2$, respectively. Here, if we let $\hat{y}_i = x_{i,1}\hat{\beta}_1 + x_{i,2}\hat{\beta}_2$, we have

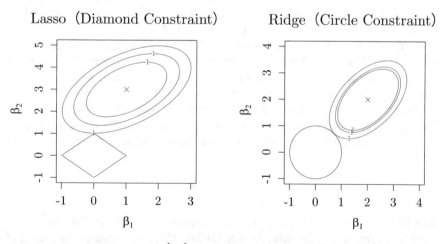

Fig. 1.7 Each ellipse is centered at $(\hat{\beta}_1, \hat{\beta}_2)$, representing the contour line connecting all the points that give the same value of (1.13). The rhombus in the left figure is the L1 regularization constraint $|\beta_1| + |\beta_2| \leq C'$, while the circle in the right figure is the L2 regularization constraint $\beta_1^2 + \beta_2^2 \leq C$

$$\sum_{i=1}^{N} x_{i,1}(y_i - \hat{y}_i) = \sum_{i=1}^{N} x_{i,2}(y_i - \hat{y}_i) = 0.$$

However, since for any β_1, β_2,

$$y_i - \beta_1 x_{i,1} - \beta_2 x_{i,2} = y_i - \hat{y}_i - (\beta_1 - \hat{\beta}_1)x_{i,1} - (\beta_2 - \hat{\beta}_2)x_{i,2}$$

holds, we can rewrite the quantity to be minimized, $\sum_{i=1}^{N}(y_i - \beta_1 x_{i,1} - \beta_2 x_{i,2})^2$, as

$$(\beta_1 - \hat{\beta}_1)^2 \sum_{i=1}^{N} x_{i,1}^2 + 2(\beta_1 - \hat{\beta}_1)(\beta_2 - \hat{\beta}_2) \sum_{i=1}^{N} x_{i,1} x_{i,2} + (\beta_2 - \hat{\beta}_2)^2 \sum_{i=1}^{N} x_{i,2}^2 + \sum_{i=1}^{N}(y_i - \hat{y}_i)^2$$

(1.13)

and, of course, if we let $(\beta_1, \beta_2) = (\hat{\beta}_1, \hat{\beta}_2)$ here, we obtain the minimum (= RSS).

Thus, we can view the problems of Lasso and Ridge in the following way: the minimization of quantities (1.5), (1.6) is equivalent to finding the values of (β_1, β_2) that satisfy the constraints $\beta_1^2 + \beta_2^2 \leq C$ and $|\beta_1| + |\beta_2| \leq C'$, respectively, that also minimize the quantity of (1.13) (here, the case where λ is large is equivalent to the case where C, C' are small).

The case of Lasso is the same as in the left panel of Fig. 1.7. The ellipses are centered at $(\hat{\beta}_1, \hat{\beta}_2)$ and represent contours on which the values of (1.13) are the same. We expand the size of the ellipse (the contour), and once we make contact with the rhombus, the corresponding values of (β_1, β_2) are the solution to Lasso. If the rhombus is small (λ is large), it is more likely to touch only one of the four rhombus vertices. In this case, one of the β_1, β_2 values will become 0. However, in the Ridge case, as in the right panel of Fig. 1.7, a circle replaces the Lasso rhombus; hence, it is less likely that $\beta_1 = 0, \beta_2 = 0$ will occur.

In this case, if the least squares solution $(\hat{\beta}_1, \hat{\beta}_2)$ lies in the green zone of Fig. 1.8, then we have either $\beta_1 = 0$ or $\beta_2 = 0$ as our solution. Moreover, when the rhombus is small (λ is large), even when $(\hat{\beta}_1, \hat{\beta}_2)$ remains the same, the green zone will become larger, which is the reason why Lasso performs well in variable selection.

We should not overlook one of the advantages of Ridge: its performance when dealing with the case of collinearity. That is, it can handle well even the case where the matrix of explanatory variables contains columns that are highly related. Let us define by

$$VIF := \frac{1}{1 - R_{X_j|X_{-j}}^2}$$

the VIF (variance inflation factor). The larger this value is, the better the j-th column variable is explained by the other variables. Here, $R_{X_j|X_{-j}}^2$ denotes the coefficient of determination squared in which X_j is the target variable, and the other variables are predictors.

Fig. 1.8 The green zone represents the area in which the optimal solution would satisfy either $\beta_1 = 0$, or $\beta_2 = 0$ if the center of the ellipses $(\hat{\beta}_1, \hat{\beta}_2)$ lies within it

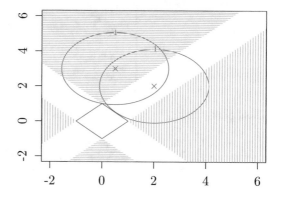

Example 6 We compute the VIF for the Boston dataset. It shows that the ninth and tenth variables (RAD and TAX) have strong collinearity.

```
1  R2 = function(x, y) {
2    y.hat = lm(y ~ x)$fitted.values; y.bar = mean(y)
3    RSS = sum((y - y.hat) ^ 2); TSS = sum((y - y.bar) ^ 2)
4    return(1 - RSS / TSS)
5  }
6  vif = function(x) {
7    p = ncol(x); values = array(dim = p)
8    for (j in 1:p) values[j] = 1 / (1 - R2(x[, -j], x[, j]))
9    return(values)
10 }
11 library(MASS); x = as.matrix(Boston); vif(x)
```

```
[1] 1.831537 2.352186 3.992503 1.095223 4.586920 2.260374 3.100843
[8] 4.396007 7.808198 9.205542 1.993016 1.381463 3.581585 3.855684
```

In usual linear regression, if the VIF is large, the estimated coefficient $\hat{\beta}$ will be unstable. Particularly, if two columns are precisely the same, the coefficient is unsolvable. Moreover, for Lasso, if two columns are highly related, then generally one of them will be estimated as 0 and the other as nonzero. However, in the Ridge case, for $\lambda > 0$, even when the columns j, k of X are the same, the estimation is solvable, and both of them will obtain the same value.

In particular, we find the partial derivative of

$$L = \frac{1}{N} \sum_{i=1}^{N} (y_i - \sum_{j=1}^{p} x_{i,j}\beta_j)^2 + \lambda \sum_{j=1}^{p} \beta_j^2$$

with respect to β_k, β_l and make it equal to 0.

$$0 = \begin{cases} -\dfrac{1}{N} \sum_{i=1}^{N} x_{i,k}(y_i - \sum_{j=1}^{p} x_{i,j}\beta_j) + \lambda\beta_k \\[3mm] -\dfrac{1}{N} \sum_{i=1}^{N} x_{i,l}(y_i - \sum_{j=1}^{p} x_{i,j}\beta_j) + \lambda\beta_l. \end{cases}$$

Then, plugging $x_{i,k} = x_{i,l}$ into each gives

$$\beta_k = \frac{1}{\lambda N} \sum_{i=1}^{N} x_{i,k}(y_i - \sum_{j=1}^{p} x_{i,j}\beta_j) = \frac{1}{\lambda N} \sum_{i=1}^{N} x_{i,l}(y_i - \sum_{j=1}^{p} x_{i,j}\beta_j) = \beta_l.$$

Example 7 We perform Lasso for the case where the variables X_1, X_2, X_3 and X_4, X_5, X_6 have strong correlations. We generate $N = 500$ groups of data distributed in the following way:

$$z_1, z_2, \epsilon, \epsilon_1, \ldots, \epsilon_6 \sim N(0, 1)$$
$$\begin{cases} x_j := z_1 + \epsilon_j/5, \ j = 1, 2, 3 \\ x_j := z_2 + \epsilon_j/5, \ j = 4, 5, 6 \end{cases}$$
$$y := 3z_1 - 1.5z_2 + 2.\epsilon$$

Then, we apply linear regression analysis with Lasso to $X \in \mathbb{R}^{N \times p}$, $y \in \mathbb{R}^N$. We plot how the coefficients change relative to the value of λ in Fig. 1.9. Naturally, one might expect that similar coefficient values should be given to each of the related variables, though it turns out that Lasso does not behave in this way.

```
n = 500; x = array(dim = c(n, 6)); z = array(dim = c(n, 2))
for (i in 1:2) z[, i] = rnorm(n)
y = 3 * z[, 1] - 1.5 * z[, 2] + 2 * rnorm(n)
for (j in 1:3) x[, j] = z[, 1] + rnorm(n) / 5
for (j in 4:6) x[, j] = z[, 2] + rnorm(n) / 5
glm.fit = glmnet(x, y); plot(glm.fit)
legend("topleft", legend = c("X1", "X2", "X3", "X4", "X5", "X6"), col = 1:6,
    lwd = 2, cex = .8)
```

1.6 Elastic Net

Up until now, we have discussed the pros and cons of Lasso and Ridge. This section studies a method intended to combine the advantages of the two, i.e., the elastic net. Specifically, the method considers the problem of finding β that minimizes

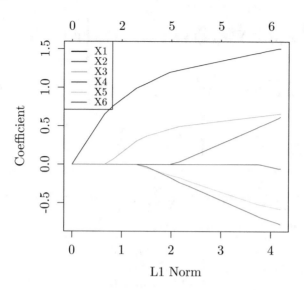

Fig. 1.9 The execution result of Example 7. The Lasso case is different from that of Ridge: related variables are not given similar estimated coefficients. When we use the `glmnet` package, its default horizontal axis is the L1-norm [11]. This is the value $\sum_{j=1}^{p} |\beta_j|$, which becomes smaller as λ increases. Therefore, the figure here is left-/right-reversed compared to that where λ or $\log \lambda$ is set as the horizontal axis.

$$L := \frac{1}{2N} \|y - X\beta\|_2^2 + \lambda \left\{ \frac{1-\alpha}{2} \|\beta\|_2^2 + \alpha \|\beta\|_1 \right\}. \tag{1.14}$$

The β_j that minimizes (1.14) for the case of Lasso ($\alpha = 1$) is $\hat{\beta}_j = \dfrac{S_\lambda(\frac{1}{N} \sum_{i=1}^{N} r_{i,j} x_{i,j})}{\frac{1}{N} \sum_{i=1}^{N} x_{i,j}^2}$,

while for the case of Ridge ($\alpha = 0$), it is $\hat{\beta}_j = \dfrac{\frac{1}{N} \sum_{i=1}^{N} r_{i,j} x_{i,j}}{\frac{1}{N} \sum_{i=1}^{N} x_{i,j}^2 + \lambda}$. In general cases, it is

$$\hat{\beta}_j = \frac{S_{\lambda \alpha}(\frac{1}{N} \sum_{i=1}^{N} r_{i,j} x_{i,j})}{\frac{1}{N} \sum_{i=1}^{N} x_{i,j}^2 + \lambda(1 - \alpha)}. \tag{1.15}$$

This is called the elastic net. For each $j = 1, \ldots, p$, if we find the subderivative of (1.14) with respect to β_j, we obtain

$$0 \in -\frac{1}{N}\sum_{i=1}^{N} x_{i,j}(y_i - \sum_{k=1}^{p} x_{i,k}\beta_k) + \lambda(1-\alpha)\beta_j + \lambda\alpha \begin{cases} 1, & \beta_j > 0 \\ [-1,1], & \beta_j = 0 \\ -1, & \beta_j < 0 \end{cases}$$

$$\iff 0 \in -\frac{1}{N}\sum_{i=1}^{N} x_{i,j}r_{i,j} + \left\{\frac{1}{N}\sum_{i=1}^{N} x_{i,j}^2 + \lambda(1-\alpha)\right\}\beta_j + \lambda\alpha \begin{cases} 1, & \beta_j > 0 \\ [-1,1], & \beta_j = 0 \\ -1, & \beta_j < 0 \end{cases}$$

$$\iff \left\{\frac{1}{N}\sum_{i=1}^{N} x_{i,j}^2 + \lambda(1-\alpha)\right\}\beta = \mathcal{S}_{\lambda\alpha}\left(\frac{1}{N}\sum_{i=1}^{N} x_{i,j}r_{i,j}\right).$$

Here, let $s_j := \frac{1}{N}\sum_{i=1}^{N} x_{i,j}r_{i,j}$, $t_j := \frac{1}{N}\sum_{i=1}^{N} x_{i,j}^2 + \lambda(1-\alpha)$, and $\mu := \lambda\alpha$. We have used the fact that

$$0 \in -s_j + t_j\beta_j + \mu \begin{cases} 1, & \beta_j > 0 \\ [-1,1], & \beta_j = 0 \\ -1, & \beta_j < 0 \end{cases} \iff t_j\beta_j = \mathcal{S}_\mu(s_j),$$

where $\mathcal{S}_\lambda : \mathbb{R}^p \to \mathbb{R}^p$ is a function that returns $\mathcal{S}_\lambda$ for each element.

Then, we can write the program for the elastic net based on (1.15), as shown in the following. Here, we added a parameter α to the line of #, and the only essential change lies in the three lines of ##.

```
linear.lasso = function(X, y, lambda = 0, beta = rep(0, ncol(X)), alpha = 1)
   { #
   X = as.matrix(X); n = nrow(X); p = ncol(X); X.bar = array(dim = p)
   for (j in 1:p) {X.bar[j] = mean(X[, j]); X[, j] = X[, j] - X.bar[j]}
   y.bar = mean(y); y = y - y.bar
   scale = array(dim = p)
   for (j in 1:p) {scale[j] = sqrt(sum(X[, j] ^ 2) / n); X[, j] = X[, j] /
     scale[j]}
   eps = 1; beta.old = beta
   while (eps > 0.001) {
     for (j in 1:p) {
       r = y - as.matrix(X[, -j]) %*% beta[-j]
       beta[j] = soft.th(lambda * alpha,                 ##
       sum(r * X[, j]) / n) / (sum(X[, j] * X[, j]) / n + ##
         lambda * (1 - alpha))                           ##
     }
     eps = max(abs(beta - beta.old)); beta.old = beta
   }
   for (j in 1:p) beta[j] = beta[j] / scale[j]
   beta.0 = y.bar - sum(X.bar * beta)
   return(list(beta = beta, beta.0 = beta.0))
}
```

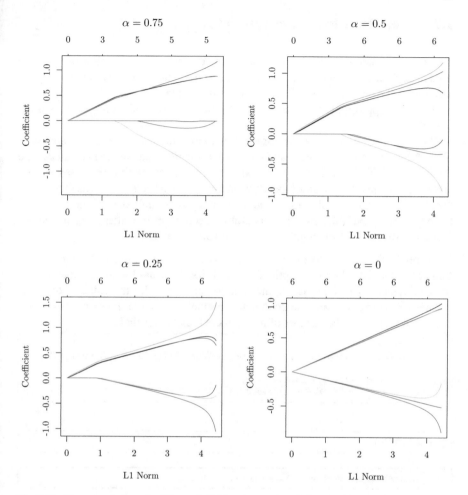

Fig. 1.10 The execution result of Example 8. The closer α is to 0 (the closer the model is to Ridge), the more it is able to handle collinearity, which is in contrast to the case of $\alpha = 1$ (Lasso), where the related variable coefficients are not estimated equally

Example 8 If we add the additional parameter `alpha = 0, 0.25, 0.5, 0.75` to `glm.fit = glmnet(x, y)` of Example 7, we obtain the graph of Fig. 1.10. As α approaches 0, we can observe how the coefficients of related variables become closer to one another. This outcome reveals how Ridge responds to collinearity.

1.7   About How to Set the Value of λ

To perform Lasso, the CRAN package `glmnet` is often used. Until now, we have tried to explain the theory behind calculations from scratch; however, it is also a good idea to use the precompiled function from said package.

To set an appropriate value for λ, the method of cross-validation (CV) is often used.[3]

For example, the tenfold CV for each λ divides the data into ten groups, with nine of them used to estimate β and one used as the test data, and then evaluates the model. Switching the test group, we can perform this evaluation ten times in total and then calculate the mean of these evaluation values. Then, we choose the λ that has the highest mean evaluation value. If we plug the sample data of the target and the explanatory variables into the function `cv.glmnet`, it evaluates each value of λ and returns the λ of the highest evaluation value as an output.

Example 9 We apply the function `cv.glmnet` to the U.S. crime dataset from Examples 2 and 5, obtain the optimal λ, use it for the usual Lasso, and then obtain the coefficients of β. For each λ, the function also provides the value of the least squares of the test data and the confidence interval (Fig. 1.11). Each number above the figure represents the number of nonzero coefficients for that λ.

```
library(glmnet)
df = read.table("crime.txt"); X = as.matrix(df[, 3:7]); y = as.vector(df[,
    1])
cv.fit = cv.glmnet(X, y); plot(cv.fit)
lambda.min = cv.fit$lambda.min; lambda.min
```

```
[1] 20.03869
```

```
fit = glmnet(X, y, lambda = lambda.min); fit$beta
```

```
5 x 1 sparse Matrix of class "dgCMatrix"
          s0
V3  9.656911
V4 -2.527286
V5  3.229431
V6  .
V7  .
```

For the elastic net, we have to perform double loop cross-validation for both (α, λ). The function `cv.glmnet` provides us the output `cvm`, which contains the evaluation values from the cross-validation.

---

[3] Please refer to Chap. 3 of "100 Problems in Mathematics of Statistical Machine Learning with R" of the same series.

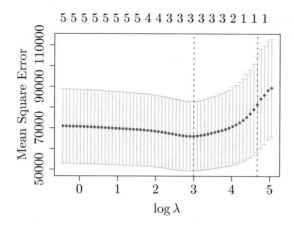

Fig. 1.11 The execution result of Example 9. Using `cv.glmnet`, we obtain the evaluation value (the sum of the squared residuals from the test data) for each λ. This is represented by the red dots in the figure. Moreover, the extended lines above and below are confidence intervals for the actual values. The optimal value here is approximately $\log \lambda_{min} = 3$ ($\lambda_{min} = 20$). Each number $5, \ldots, 5, 4, 4, 3, \ldots, 3, 2, 1, 1, 1$ at the top represents the number of nonzero coefficients

Example 10 We generate random numbers as our data and try conducting the double loop cross-validation for (α, λ).

```
1  n = 500; x = array(dim = c(n, 6)); z = array(dim = c(n, 2))
2  for (i in 1:2) z[, i] = rnorm(n)
3  y = 3 * z[, 1] - 1.5 * z[, 2] + 2 * rnorm(n)
4  for (j in 1:3) x[, j] = z[, 1] + rnorm(n) / 5
5  for (j in 4:6) x[, j] = z[, 2] + rnorm(n) / 5
6  best.score = Inf
7  for (alpha in seq(0, 1, 0.01)) {
8    res = cv.glmnet(x, y, alpha = alpha)
9    lambda = res$lambda.min; min.cvm = min(res$cvm)
10   if (min.cvm < best.score) {alpha.min = alpha; lambda.min = lambda; best.
         score = min.cvm}
11 }
12 alpha.min
```

```
[1] 0.47
```

```
1  lambda.min
```

```
[1] 0.05042894
```

```
1  glmnet(x, y, alpha = alpha.min, lambda = lambda.min)$beta
```

Fig. 1.12 Estimates of the
coefficients of the six
variables from Example 7.
Here, we use the
cross-validated optimal α as
the parameter

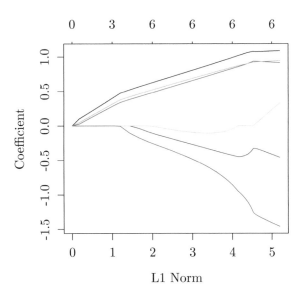

```
1   6 x 1 sparse Matrix of class "dgCMatrix"
2            s0
3   V1   1.0562856
4   V2   0.9382231
5   V3   0.9258483
6   V4  -0.3443867
7   V5   .
8   V6   .
```

```
1   glm.fit = glmnet(x, y, alpha = alpha.min)
2   plot(glm.fit)
```

We plot the graph of the coefficient values for when α is optimal (in the cross-validation sense) in Fig. 1.12.

Exercises 1–20

In the following exercises, we estimate the intercept β_0 and the coefficients $\beta = (\beta_1, \ldots, \beta_p)$ from N groups of p explanatory variables and a target variable data $(x_{1,1}, \ldots, x_{1,p}, y_1), \ldots, (x_{N,1}, \ldots, x_{N,p}, y_N)$. We subtract from each $x_{i,j}$ the $\bar{x}_j = \frac{1}{N} \sum_{i=1}^{N} x_{i,j}$ $(j = 1, \ldots, p)$ and from each y_i the $\bar{y} = \frac{1}{N} \sum_{i=1}^{N} y_i$ such that each of them has mean 0 and then estimate β (the estimated value is denoted by $\hat{\beta}$). Finally, we let $\hat{\beta}_0 = \bar{y} - \sum_{j=1}^{p} \bar{x}_j \hat{\beta}_j$. Again, we let $X = (x_{i,j})_{1 \leq i \leq N, 1 \leq j \leq p} \in \mathbb{R}^{N \times p}$ and $y = (y_i)_{1 \leq i \leq N} \in \mathbb{R}^N$.

1. Prove that the following two equalities hold:

$$
\begin{bmatrix}
\dfrac{\partial}{\partial \beta_1} \displaystyle\sum_{i=1}^{N}(y_i - \sum_{k=1}^{p}\beta_k x_{i,k})^2 \\
\vdots \\
\dfrac{\partial}{\partial \beta_p} \displaystyle\sum_{i=1}^{N}(y_i - \sum_{k=1}^{p}\beta_k x_{i,k})^2
\end{bmatrix}
= -2
\begin{bmatrix}
x_{1,1} & \cdots & x_{N,1} \\
\vdots & \ddots & \vdots \\
x_{1,p} & \cdots & x_{N,p}
\end{bmatrix}
\begin{bmatrix}
y_1 - \sum_{k=1}^{p}\beta_k x_{1,k} \\
\vdots \\
y_N - \sum_{k=1}^{p}\beta_k x_{N,k}
\end{bmatrix}
$$

$$
= -2X^T(y - X\beta).\text{(cf. (1.1))}
$$

Moreover, when $X^T X$ is invertible, prove that the value of β that minimizes $\|y - X\beta\|_2^2 := \sum_{i=1}^{N}(y_i - \sum_{k=1}^{p}\beta_1 x_{i,k})^2$ is given by $\hat{\beta} = (X^T X)^{-1} X^T y$. Here, for $z = [z_1, \ldots, z_N]^T \in \mathbb{R}^N$, we define $\|z\|_2 := \sqrt{z_1^2 + \cdots + z_N^2}$. Then, write a program for a function called `linear` with the R language. This function accepts a matrix $X \in \mathbb{R}^{N \times P}$ and a vector $y \in \mathbb{R}^N$ as inputs and then calculates the estimated intercept $\hat{\beta}_0 \in \mathbb{R}$ and the estimated slope $\hat{\beta} \in \mathbb{R}^p$ as outputs. Please complete blanks (1), (2) below. Here, the function `centralize` conducting mean centering to X, y is used. In the third and fourth lines below, X, y has already been centered.

```
1  linear = function(X, y) {
2      n = nrow(X); p = ncol(X)
3      res = centralize(X, y, standardize = FALSE)
4      X = res$X; y = res$y
5      beta = as.vector(## blank (1) ##)
6      beta.0 = ## blank (2) ##
7      return(list(beta = beta, beta.0 = beta.0))
8  }
```

2. For a function $f : \mathbb{R} \to \mathbb{R}$ that is convex (downward), for $x_0 \in \mathbb{R}$, we have

$$
f(x) \geq f(x_0) + z(x - x_0) \quad (x \in \mathbb{R}). \qquad \text{(cf. (1.8)}
$$

The set of $z \in \mathbb{R}$ (subderivative) is denoted by $\partial f(x_0)$. Show that the following function f is convex. In addition, at $x_0 = 0$, find $\partial f(x_0)$.

(a) $f(x) = x$

(b) $f(x) = |x|$

Hint For any $0 < \alpha < 1$ and $x, y \in \mathbb{R}$, if $f(\alpha x + (1 - \alpha)y) \leq \alpha f(x) + (1 - \alpha)f(y)$ holds, then we say that f is convex. For (b), show that $|x| \geq zx$ $(x \in \mathbb{R}) \iff -1 \leq z \leq 1$.

3. (a) If the functions $g(x), h(x)$ are convex, show that for any $\beta, \gamma \geq 0$, the function $\beta g(x) + \gamma h(x)$ is convex.

(b) Show that the function $f(x) = \begin{cases} 1, & x \neq 0 \\ 0, & x = 0 \end{cases}$ is not convex.

(c) Decide whether the following functions of $\beta \in \mathbb{R}^p$ are convex or not. In addition, provide the proofs.

i. $\dfrac{1}{2N}\|y - X\beta\|^2 + \lambda\|\beta\|_0$

ii. $\dfrac{1}{2N}\|y - X\beta\|^2 + \lambda\|\beta\|_1$

iii. $\dfrac{1}{N}\|y - X\beta\|^2 + \lambda\|\beta\|_2^2$

Here, $\|\cdot\|_0$ denotes the number of nonzero elements, $\|\cdot\|_1$ denotes the sum of the absolute values of all elements, and $\|\cdot\|_2$ denotes the square root of the sum of the square of each element. Moreover, let $\lambda \geq 0$.

4. For a convex function $f : \mathbb{R} \to \mathbb{R}$, assume that it is differentiable at $x = x_0$. Answer the following questions.

(a) Show that $f(x) \geq f(x_0) + f'(x_0)(x - x_0)$.
Hint From $f(\alpha x + (1 - \alpha)x_0) \leq \alpha f(x) + (1 - \alpha)f(x_0)$, derive the following formula and take the limit.

$$f(x) \geq f(x_0) + \frac{f(x_0 + \alpha(x - x_0)) - f(x)}{\alpha(x - x_0)}(x - x_0).$$

Show that the set
(b) $\partial f(x_0)$ contains only the derivative of f at $x = x_0$ as its element.
Hint In the last part, you may carry out a calculation similar to the following:

$$f(x) \geq f(x_0) + z(x - x_0) \quad (x \in \mathbb{R})$$

$$\implies \begin{cases} \dfrac{f(x) - f(x_0)}{x - x_0} \geq z, & x > x_0 \\ \dfrac{f(x) - f(x_0)}{x - x_0} \leq z, & x < x_0 \end{cases}$$

$$\implies \lim_{x \to x_0 - 0} \frac{f(x) - f(x_0)}{x - x_0} \leq z \leq \lim_{x \to x_0 + 0} \frac{f(x) - f(x_0)}{x - x_0}.$$

However, this case still leaves two possibilities: either $\{f'(x_0)\}$ or an empty set.

5. Find the local minimum of the following functions. In addition, write the graphs of the range $-2 \leq x \leq 2$ using the R language.

(a) $f(x) = x^2 - 3x + |x|$ and
(b) $f(x) = x^2 + x + 2|x|$.
Hint Divide it into three cases: $x < 0$, $x = 0$, $x > 0$.

6. The (i, j) entry of $x \in \mathbb{R}^{N \times p}$ and the i-th element of $y \in \mathbb{R}^N$ are denoted by $x_{i,j}$ and y_i, respectively. Let $\lambda \geq 0$; we want to find $\beta_1, \ldots, \beta_p$ that minimize

$$L := \frac{1}{2N} \sum_{i=1}^{N} (y_i - \sum_{j=1}^{p} \beta_j x_{i,j})^2 + \lambda \sum_{j=1}^{p} |\beta_j|.$$

Here, we assume that

$$\frac{1}{N} \sum_{i=1}^{N} x_{i,j} x_{i,k} = \begin{cases} 1, & j = k \\ 0, & j \neq k \end{cases} \tag{1.16}$$

and let $s_j = \frac{1}{N} \sum_{i=1}^{N} x_{i,j} y_i$. For $\lambda > 0$, holding β_k ($k \neq j$) constant, find the sub-derivative of L with respect to β_j. In addition, find the value of β_j that attains the minimum.

Hint Consider each of the following cases: $\beta_j > 0$, $\beta_j = 0$, $\beta_j < 0$. For example, when $\beta_j = 0$, we have $-\beta_j + s_j - \lambda[-1, 1] \ni 0$, but this is equivalent to $-\lambda \leq s_j \leq \lambda$.

7. (a) Let $\lambda \geq 0$; define a function $S_\lambda(x)$ as follows:

$$S_\lambda(x) := \begin{cases} x - \lambda, & x > \lambda \\ 0, & |x| \leq \lambda \\ x + \lambda, & x < -\lambda. \end{cases}$$

For this function $S_\lambda(x)$, by using the function $(x)_+ = \max\{x, 0\}$ and

$$\text{sign}(x) = \begin{cases} -1, & x < 0 \\ 0, & x = 0 \\ 1, & x > 0 \end{cases}$$

show that it can be re-expressed as $S_\lambda(x) = \text{sign}(x)(|x| - \lambda)_+$.

(b) Write a program in the R language that performs (a). Name it `soft.th`. Then, execute the following code to see whether it works correctly.

```
curve(soft.th(5, x), -10, 10)
```

Hint In the R language, if we define a function by dividing it into cases or using `max`, it will not plot a graph for us. Use `sign`, `abs`, and `pmax` instead.

8. For the cases where the assumption (1.16) of Exercise 6 is not satisfied, if we fix β_k ($k \neq j$) and then update

$$\begin{cases} r_i := y_i - \sum_{k \neq j} \beta_k x_{i,k} & (i = 1, \dots, n) \\ \beta_j := S_\lambda \left(\frac{1}{n} \sum_{i=1}^{n} x_{i,j} r_i \right) \Big/ \left(\frac{1}{n} \sum_{i=1}^{n} x_{i,j}^2 \right) \end{cases}$$

for $j = 1, \ldots, p$ sequentially, by repeating the process, we can solve for $\beta_1, \ldots,$ β_p. Moreover, to calculate the intercept, initially center the data; then, the intercept can be calculated by letting

$$\beta_0 := \bar{y} - \sum_{j=1}^{p} \beta_j \bar{x}_j.$$

Here, we denote $\bar{x}_j = \dfrac{1}{N} \sum_{i=1}^{n} x_{i,j}$ and $\bar{y}_j = \dfrac{1}{N} \sum_{i=1}^{n} y_i$. Complete the blanks in the following and execute the program. In addition, by letting $\lambda = 10, 50, 100$, check how the coefficients that once were 0 change accordingly.

To execute the following codes, please access the site https://web.stanford.edu/~hastie/StatLearnSparsity/data.html and download the U.S. crime data.[4] Then, store them in a file named crime.txt.[5]

```
linear.lasso = function(X, y, lambda = 0) {
  X = as.matrix(X); n = nrow(X); p = ncol(X)
  res = centralize(X, y); X = res$X; y = res$y
  eps = 1; beta = rep(0, p); beta.old = rep(0, p)
  while (eps > 0.001) {
    for (j in 1:p) {
      r = ## blank (1) ##
      beta[j] = soft.th(lambda, sum(r * X[, j]) / n)
    }
    eps = max(abs(beta - beta.old)); beta.old = beta
  }
  beta = beta / res$X.sd
  beta.0 = ## blank (2) ##
  return(list(beta = beta, beta.0 = beta.0))
}
crime = read.table("crime.txt"); X = crime[, 3:7]; y = crime[, 1]
linear.lasso(X, y, 10); linear.lasso(X, y, 50); linear.lasso(X, y, 100)
```

9. `glmnet` is an R package commonly used to perform Lasso. The following code performs the regression analysis with sparsity to the U.S. crime data using `glmnet`.

```
library(glmnet)
df = read.table("crime.txt"); x = as.matrix(df[, 3:7]); y = df[, 1]
fit = glmnet(x, y); plot(fit, xvar = "lambda")
```

---

[4] Open the window, press Ctrl+A to select all, press Ctrl+C to copy, make a new file of your text editor, and then press Ctrl+V to paste the data.

[5] To execute the R program, you may need to set the working directory or provide the path to where the dataset file is located.

Following the above, conduct an analysis similar to that of the Boston dataset[6] by setting the 14th column as the target variable and the other 13 as explanatory and then write a graph in the same manner.

Column	Variable	Definition of Variable
1	CRIM	per capita crime rate by town
2	ZN	proportion of residential land zoned for lots over 25,000 sq.ft.
3	INDUS	proportion of nonretail business acres per town
4	CHAS	Charles River dummy variable (1 if tract bounds river; 0 otherwise)
5	NOX	nitric oxide concentration (parts per 10 million)
6	RM	average number of rooms per dwelling
7	AGE	proportion of owner-occupied units built prior to 1940
8	DIS	weighted distances to five Boston employment centers
9	RAD	index of accessibility to radial highways
10	TAX	full-value property-tax rate per $10,000
11	PTRATIO	pupil-teacher ratio by town
12	BLACK	proportion of blacks by town
13	LSTAT	% lower status of the population
14	MEDV	median value of owner-occupied homes in multiples of $1,000

10. The coordinate descent starts from a large λ so that all coefficients are initially 0 and then gradually decreases the size of λ. In addition, for each iteration, consider using the coefficients estimated with the last λ as the initial values for the next λ iteration (warm start). We consider the following program and apply it to U.S. crime data. Fill in the blank.

```
1  warm.start = function(X, y, lambda.max = 100) {
2    dec = round(lambda.max / 50); lambda.seq = seq(lambda.max, 1, -dec)
3    r = length(lambda.seq); p = ncol(X); coef.seq = matrix(nrow = r, ncol =
        p)
4    coef.seq[1, ] = linear.lasso(X, y, lambda.seq[1])$beta
5    for (k in 2:r) coef.seq[k, ] = ## blank ##
6    return(coef.seq)
7  }
8  crime = read.table("crime.txt"); X = crime[, 3:7]; y = crime[, 1]
9  coef.seq = warm.start(X, y, 200)
10 p = ncol(X)
11 lambda.max = 200
12 dec = round(lambda.max / 50)
13 lambda.seq = seq(lambda.max, 1, -dec)
14 plot(log(lambda.seq), coef.seq[, 1], xlab = "log(lambda)", ylab = "
        Coefficient",
15       ylim = c(min(coef.seq), max(coef.seq)), type = "n")
16 for (j in 1:p) lines(log(lambda.seq), coef.seq[, j], col = j)
```

Next, setting λ large enough for all initial coefficients to be 0, we perform the coordinate descent. For each j, show that when both $\sum_{i=1}^{N} x_{i,j}^2 = 1$ and $\sum_{i=1}^{N} x_{i,j} y_i$ are different for all $j = 1, \ldots, p$, the smallest λ that makes every

---

[6] Please install the MASS package.

variable coefficient 0 is given by $\lambda = \max\limits_{1 \le j \le p} \left| \dfrac{1}{N} \sum\limits_{i=1}^{N} x_{i,j} y_i \right|$. Moreover, execute the following code:

```
X = as.matrix(X); y = as.vector(y); cv = cv.glmnet(X, y); plot(cv)
```

What is the meaning of the number n?

11. Denote the eigenvalues of $X^T X$ by $\gamma_1, \ldots, \gamma_p$. Using them, derive the condition where the inverse of $X^T X$ does not exist. In addition, show that the eigenvalues of $X^T X + N\lambda I$ are given by $\gamma_1 + N\lambda, \ldots, \gamma_p + N\lambda$. Moreover, whenever $\lambda > 0$ is satisfied, show that $X^T X + N\lambda I$ always has an inverse. Here, we can use the fact, without proof, that the eigenvalues of a positive semidefinite matrix are all nonnegative.

12. Let $\lambda \ge 0$. Find the β that minimizes

$$\frac{1}{2N}\|y - X\beta\|_2^2 + \frac{\lambda}{2}\|\beta\|_2^2 := \frac{1}{2N}\sum_{i=1}^{N}(y_i - \beta_1 x_{i,1} - \cdots - \beta_p x_{i,p})^2 + \frac{\lambda}{2}\sum_{j=1}^{p}\beta_j^2 \text{ (cf. (1.8))}$$

(Ridge regression).

13. Adding to the function `linear` of Exercise 1 the additional parameter $\lambda \ge 0$, write an R program `ridge` that performs the process of Exercise 1.7. Execute that program, and check whether your results match the following examples.

```
crime = read.table("crime.txt")
X = crime[, 3:7]
y = crime[, 1]
linear(X, y)
```

```
$beta
[1] 10.9806703 -6.0885294  5.4803042  0.3770443  5.5004712
$beta.0
[1] 489.6486
```

The output of the function `ridge` should be as follows:

```
ridge(X, y)
```

```
$beta
[1] 10.9806703 -6.0885294  5.4803042  0.3770443  5.5004712
$beta.0
[1] 489.6486
```

```
1 ridge(X, y, 200)
```

```
1   $beta
2   [1] 0.056351799 -0.019763974   0.077863094 -0.017121797 -0.007039304
3   $beta.0
4   [1] 716.4044
```

14. The code below performs an analysis of the U.S. crime data in the following order: while changing the value of λ, it solves for coefficients using the function ridge and plots a graph of how each coefficient changes. In the program, change lambda.seq to log(lambda.seq), and change the horizontal axis label from lambda to log(lambda). Then, plot the graph (and change the title (the main part)).

```
1  df = read.table("crime.txt"); x = df[, 3:7]; y = df[, 1]; p = ncol(x)
2  lambda.max = 3000; lambda.seq = seq(1, lambda.max)
3  plot(lambda.seq, xlim = c(0, lambda.max), ylim = c(-12, 12),
4       xlab = "lambda", ylab = "coefficients", main = "changes in
       coefficients according to lambda",
5       type = "n", col = "red")                 ## this 1 line
6  for (j in 1:p) {
7    coef.seq = NULL
8    for (lambda in lambda.seq) coef.seq = c(coef.seq, ridge(x, y, lambda)$
       beta[j])
9    par(new = TRUE)
10   lines(lambda.seq, coef.seq, col = j)        ## this 1 line
11 }
12 legend("topright",
13        legend = c("annual police funding", "\% of people 25 years+ with 4
       yrs. of high school",
14                   "\% of 16--19 year-olds not in highschool and not
       highschool graduates",
15                   "\% of people 25 years+ with at least 4 years of
       college"),
16        col = 1:p, lwd = 2, cex = .8)
```

15. For given $x_{i,1}, x_{i,2}, y_i \in \mathbb{R}$ ($i = 1, \ldots, N$), the β_1, β_2 that minimize $S := \sum_{i=1}^{N} (y_i - \beta_1 x_{i,1} - \beta_2 x_{i,2})^2$ is denoted by $\hat{\beta}_1, \hat{\beta}_2$, and $\hat{\beta}_1 x_{i,1} + \hat{\beta}_2 x_{i,2}$ is denoted by $\hat{y}_i$, ($i = 1, \ldots, N$).

(a) Prove that the following two equalities hold:

i. $\displaystyle\sum_{i=1}^{N} x_{i,1}(y_i - \hat{y}_i) = \sum_{i=1}^{N} x_{i,2}(y_i - \hat{y}_i) = 0.$

ii. For any β_1, β_2,

$$y_i - \beta_1 x_{i,1} - \beta_2 x_{i,2} = y_i - \hat{y}_i - (\beta_1 - \hat{\beta}_1)x_{i,1} - (\beta_2 - \hat{\beta}_2)x_{i,2}.$$

In addition, for any β_1, β_2, show that the quantity $\sum_{i=1}^{N}(y_i - \beta_1 x_{i,1} - \beta_2 x_{i,2})^2$ can be rewritten as

$$(\beta_1 - \hat{\beta}_1)^2 \sum_{i=1}^{N} x_{i,1}^2 + 2(\beta_1 - \hat{\beta}_1)(\beta_2 - \hat{\beta}_2) \sum_{i=1}^{N} x_{i,1}x_{i,2} + (\beta_2 - \hat{\beta}_2)^2 \sum_{i=1}^{N} x_{i,2}^2$$

$$+ \sum_{i=1}^{N}(y_i - \hat{y}_i)^2 . \text{(cf. (1.13))}$$

(b) Assume that $\sum_{i=1}^{N} x_{i,1}^2 = \sum_{i=1}^{N} x_{i,2}^2 = 1$, $\sum_{i=1}^{N} x_{i,1}x_{i,2} = 0$. In the case of usual least squares, we choose $\beta_1 = \hat{\beta}_1$, $\beta_2 = \hat{\beta}_2$. However, when there is a constraint requiring $|\beta_1| + |\beta_2|$ to be less than or equal to a value, we have to think of the problem as choosing a point (β_1, β_2) from a circle (of as small a radius as possible) centered at $(\hat{\beta}_1, \hat{\beta}_2)$, which has to lie within the square of the constraint. Fix the square whose vertices are $(1, 0), (0, 1), (-1, 0), (0, -1)$, and fix a point $(\hat{\beta}_1, \hat{\beta}_2)$ outside the square. Then, consider the circle centered at $(\hat{\beta}_1, \hat{\beta}_2)$. We expand the circle (increase its radius) until it touches the square. Now, draw the range of $(\hat{\beta}_1, \hat{\beta}_2)$ such that when the circle touches the square, one of the coordinates of the point of contact will be 0.

(c) In (b), what would happen if the square is replaced by a circle of radius 1 (circle touches circle)?

16. A matrix X is composed of p explanatory variables with N samples each. If it has two columns that are equal for all N samples, let $\lambda > 0$, perform Ridge regression, and show that the estimated coefficients of the two vectors will be equal.

   Hint Note that

$$L = \frac{1}{N} \sum_{i=1}^{N}(y_i - \beta_0 - \sum_{j=1}^{p} x_{i,j}\beta_j)^2 + \lambda \sum_{j=1}^{p} \beta_j^2.$$

Partially differentiating it with respect to β_k, β_l gives $-\dfrac{1}{N}\sum_{i=1}^{N} x_{i,k}(y_i - \beta_0 -$

$\sum_{j=1}^{p} x_{i,j}\beta_j) + \lambda\beta_k$ and $-\dfrac{1}{N}\sum_{i=1}^{N} x_{i,l}(y_i - \beta_0 - \sum_{j=1}^{p} x_{i,j}\beta_j) + \lambda\beta_l$. Both of them have to be 0. Plug $x_{i,k} = x_{i,l}$ into each of them.

17. We perform linear regression analysis with Lasso for the case where Y is the target variable and the variables X_1, X_2, X_3, and X_4, X_5, X_6 are highly correlated. We generate $N = 500$ groups of data, distributed as in the relations below, and then apply linear regression analysis with Lasso to $X \in \mathbb{R}^{N \times p}$, $y \in \mathbb{R}^N$.

$$z_1, z_2, \epsilon, \epsilon_1, \ldots, \epsilon_6 \sim N(0, 1)$$

$$\begin{cases} x_j := z_1 + \epsilon_j/5, & j = 1, 2, 3 \\ x_j := z_2 + \epsilon_j/5, & j = 4, 5, 6 \end{cases}$$

$$y := 3z_1 - 1.5z_2 + 2.\epsilon$$

Fill in the blanks (1), (2) below. Plot a graph showing how each coefficient changes with λ.

```
1 n = 500; x = array(dim = c(n, 6)); z = array(dim = c(n, 2))
2 for (i in 1:2) z[, i] = rnorm(n)
3 y = ## blank (1) ##
4 for (j in 1:3) x[, j] = z[, 1] + rnorm(n) / 5
5 for (j in 4:6) x[, j] = z[, 2] + rnorm(n) / 5
6 glm.fit = glmnet(## blank (2) ##); plot(glm.fit)
```

18. Instead of the usual Lasso or Ridge, we find the β_0, β values that minimize

$$\frac{1}{2N}\|y - \beta_0 - X\beta\|_2^2 + \lambda\left\{\frac{1-\alpha}{2}\|\beta\|_2^2 + \alpha\|\beta\|_1\right\} \quad \text{(cf. (1.14))} \quad (1.17)$$

(we mean-center the X, y data, initially let $\beta_0 = 0$, and restore its value later). The β_j that minimizes (1.14) is given by $\hat{\beta}_j = \dfrac{S_\lambda(\frac{1}{N}\sum_{i=1}^N r_{i,j}x_{i,j})}{\frac{1}{N}\sum_{i=1}^N x_{i,j}^2}$ in the case

of Lasso ($\alpha = 1$), by $\hat{\beta}_j = \dfrac{\frac{1}{N}\sum_{i=1}^N r_{i,j}x_{i,j}}{\frac{1}{N}\sum_{i=1}^N x_{i,j}^2 + \lambda}$ in the case of Ridge ($\alpha = 0$), or, in general cases, by

$$\hat{\beta}_j = \frac{S_{\lambda\alpha}(\frac{1}{N}\sum_{i=1}^N r_{i,j}x_{i,j})}{\frac{1}{N}\sum_{i=1}^N x_{i,j}^2 + \lambda(1-\alpha)} \quad \text{(cf. (1.15))} \quad (1.18)$$

(the elastic net cases). By finding the subderivative of (1.17) with respect to β_j and sequentially proving the three equations below, show that (1.18) holds. Here, x_j denotes the j-th column of X.

i. $0 \in -\dfrac{1}{N}X^T(y - X\beta) + \lambda(1 - \alpha)\beta_j + \lambda\alpha \begin{cases} 1, & \beta_j > 0 \\ [-1, 1], & \beta_j = 0 \\ -1, & \beta_j < 0; \end{cases}$

ii. $0 \in -\dfrac{1}{N}\sum_{i=1}^N x_{i,j}r_{i,j} + \left\{\dfrac{1}{N}\sum_{i=1}^N x_{i,j}^2 + \lambda(1-\alpha)\right\}\beta_j + \lambda\alpha \begin{cases} 1, & \beta_j > 0 \\ [-1, 1], & \beta_j = 0 \\ -1, & \beta_j < 0; \end{cases}$

iii. $\left\{\dfrac{1}{N}\sum_{i=1}^N x_{i,j}^2 + \lambda(1-\alpha)\right\}\beta = S_{\lambda\alpha}\left(\dfrac{1}{N}\sum_{i=1}^N x_{i,j}r_{i,j}\right).$

In addition, revise the function `linear.lasso` in Exercise 8, let its default parameter be `alpha = 1`, and generalize it by changing the formula to (1.18).

Moreover, correct the value of $1 - \alpha$ by dividing it by $\sqrt{\sum_{i=1}^{N}(y_i - \bar{y})^2/N}$ (this normalization exists in `glmnet`).

19. To the function `glm.fit = glmnet(x, y)` in Exercise 1.7, add the option `alpha = 0.3` and see the output. What are the differences between `alpha = 0` (Ridge) and `alpha = 1` (Lasso)?

20. Even for the case of the elastic net, we need to select the optimal value for α. In the program provided below, the output `cvm` of the `glmnet` function contains squared errors for each λ with respect to a specific value of α. Then, it assigns to that α the smallest value of the squared errors, which is attained by some λ, and the algorithm compares each α and returns the one with the smallest squared error. Once we obtain the optimal α, plot a figure similar to that of Exercise 1.7.

```
1 alpha = seq(0.01, 0.99, 0.01); m = length(alpha); mse = array(dim = m)
2 for (i in 1:m) {cvg = cv.glmnet(x, y, alpha = alpha[i]); mse[i] = min(cvg
     $cvm)}
3 best.alpha = alpha[which.min(mse)]; best.alpha
4 cva = cv.glmnet(x, y, alpha = best.alpha)
5 best.lambda = cva$lambda.min; best.lambda
```

Chapter 2
Generalized Linear Regression

In this chapter, we consider the so-called generalized linear regression, which includes logistic regression (binary and multiple cases), Poisson regression, and Cox regression. We can formulate these problems in terms of maximizing the likelihood and solve them by applying the Newton method: differentiate the log-likelihood by the parameters to be estimated, and solve the equation such that the differentiated value is zero.

In general, the solution of the maximum likelihood method may not have a finite value and may not converge even if the Newton method is applied. However, if we regularize the log-likelihood via Lasso, the larger the value of λ is, the more likely it converges, and we can choose the relevant variables.

2.1 Generalization of Lasso in Linear Regression

In Chap. 1, centralizing $X \in \mathbb{R}^{N \times p}$ and $y \in \mathbb{R}^N$ and eliminating the intercept $\beta_0 \in \mathbb{R}$, we obtain the β value that minimizes

$$\frac{1}{2N} \|y - X\beta\|^2 + \lambda \|\beta\|_1$$

for given $\lambda \geq 0$. In this section, letting $W \in \mathbb{R}^{N \times N}$ be nonnegative definite, we obtain the Lasso solution of the extended linear regression.

$$L_0 := \frac{1}{2N} (y - \beta_0 - X\beta)^T W (y - \beta_0 - X\beta) .$$

To this end, we centralize each column of X and y and eliminate β_0. In particular, if we centralize

© The Author(s), under exclusive license to Springer Nature Singapore Pte Ltd. 2021
J. Suzuki, *Sparse Estimation with Math and R*,
https://doi.org/10.1007/978-981-16-1446-0_2

$$
\begin{cases}
\bar{X}_k := \dfrac{\sum_{i=1}^{N}\sum_{j=1}^{N} w_{i,j} x_{j,k}}{\sum_{i=1}^{N}\sum_{j=1}^{N} w_{i,j}}, \quad k = 1, \ldots, p \\[2mm]
\bar{y} := \dfrac{\sum_{i=1}^{N}\sum_{j=1}^{N} w_{i,j} y_j}{\sum_{i=1}^{N}\sum_{j=1}^{N} w_{i,j}}
\end{cases}
$$

$$
\begin{cases}
x_{i,k} \leftarrow x_{i,k} - \bar{X}_k, \quad i = 1, \ldots, N, \ k = 1, \ldots, p \\
y_i \leftarrow y_i - \bar{y}, \qquad\quad i = 1, \ldots, N
\end{cases}
$$

for $i = 1, \ldots, N$, considering the weights $W = (w_{i,j})$, we have $\sum_{i=1}^{N}\sum_{j=1}^{N} w_{i,j} y_j = 0$ and $\sum_{i=1}^{N}\sum_{j=1}^{N} w_{i,j} x_{i,k} = 0$, which means that the solution

$$
\hat{\beta}_0 = \bar{y} - \sum_{k=1}^{p} \bar{X}_k \hat{\beta}_k \tag{2.1}
$$

of the equation obtained by adding all the rows

$$
\frac{\partial L_0}{\partial \beta_0} = -\frac{1}{N} W(y - \beta_0 - X\beta)
$$

$$
= -\frac{1}{N}
\begin{bmatrix}
\sum_{j=1}^{N} w_{1,j} y_j - \beta_0 \sum_{j=1}^{N} w_{1,j} - \sum_{j=1}^{N} w_{1,j} \sum_{k=1}^{p} x_{j,k} \beta_k \\
\vdots \\
\sum_{j=1}^{N} w_{N,j} y_j - \beta_0 \sum_{j=1}^{N} w_{N,j} - \sum_{j=1}^{N} w_{N,j} \sum_{k=1}^{p} x_{j,k} \beta_k
\end{bmatrix}
=
\begin{bmatrix} 0 \\ \vdots \\ 0 \end{bmatrix}.
$$

is zero. Once we obtain the centralization, we minimize

$$
L(\beta) := \frac{1}{2N}(y - X\beta)^T W (y - X\beta) + \lambda \|\beta\|_1 \ .
$$

If we execute the Cholesky decomposition as $W = M^T M$ and define $V := MX$, $u := My$, we may write

$$
L(\beta) = \frac{1}{2N}\|u - V\beta\|_2^2 + \lambda \|\beta\|_1 \ .
$$

From the obtained $\hat{\beta}$ and (2.1), we obtain $\hat{\beta}_0$.

We show a sample R program that realizes the procedure as follows:

```
1  W.linear.lasso = function(X, y, W, lambda = 0) {
2    n = nrow(X); p = ncol(X); X.bar = array(dim = p)
3    for (k in 1:p) {
4      X.bar[k] = sum(W %*% X[, k]) / sum(W)
5      X[, k] = X[, k] - X.bar[k]
6    }
7    y.bar = sum(W %*% y) / sum(W); y = y - y.bar
8    L = chol(W)
9    # L = sqrt(W)
```

```
10    u = as.vector(L %*% y); V = L %*% X
11    beta = linear.lasso(V, u, lambda)$beta
12    beta.0 = y.bar - sum(X.bar * beta)
13    return(c(beta.0, beta))
14  }
```

The matrix X, an input to the function W.linear.lasso, has p columns, corresponding to the number of variables. The output of the function is not a list but a vector of length $p + 1$ consisting of the intercept and slope in this order.

The Cholesky decomposition requires an $O(N^3)$ execution time for a matrix size N, and an error occurs when N is too large. However, in the problems addressed in this chapter, the matrix W is diagonal such that the diagonal elements are positive. In these cases, we may replace L = chol(W) with L = diag(sqrt(W)). Moreover, because we consider a sparse situation, we may assume that N is not so large compared with p.

2.2 Logistic Regression for Binary Values

Let Y be a random variable that takes values in $\{0, 1\}$. We assume that for each $x \in \mathbb{R}^p$ (row vector), there exist $\beta_0 \in \mathbb{R}$ and $\beta = [\beta_1, \ldots, \beta_p] \in \mathbb{R}^p$ such that the probability $P(Y = 1 \mid x)$ satisfies

$$\log \frac{P(Y = 1 \mid x)}{P(Y = 0 \mid x)} = \beta_0 + x\beta \tag{2.2}$$

(logistic regression). Then, we may write (2.2) as

$$P(Y = 1 \mid x) = \frac{\exp(\beta_0 + x\beta)}{1 + \exp(\beta_0 + x\beta)} . \tag{2.3}$$

Example 11 Let $p = 1$ and $\beta_0 = 0$. We display the shapes of the distributions that are given by the right-hand side of (2.3) for various $\beta \in \mathbb{R}$. The larger the value of $\beta > 0$, the more suddenly $P(Y = 1 \mid x)$ grows at $x = 0$ from almost zero to almost one. We show the curves in 2.1.

```
1   f = function(x) return(exp(beta.0 + beta * x) / (1 + exp(beta.0 + beta
          * x)))
2   beta.0 = 0; beta.seq = c(0, 0.2, 0.5, 1, 2, 10)
3   m = length(beta.seq)
4   beta = beta.seq[1]
5   plot(f, xlim = c(-10, 10), ylim = c(0, 1), xlab = "x", ylab = "y",
6        col = 1, main = "Logistic Curve")
7   for (i in 2:m) {
8     beta = beta.seq[i]
9     par(new = TRUE)
10    plot(f, xlim = c(-10, 10), ylim = c(0, 1), xlab = "", ylab = "",
          axes = FALSE, col = i)
```

Fig. 2.1 Execution of
Example 11. Setting $\beta_0 = 0$,
we draw the curves of (2.3)
when $\beta > 0$ changes

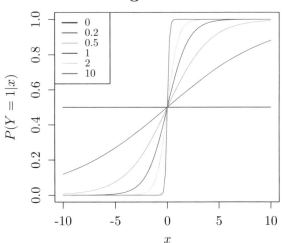

```
11  }
12  legend("topleft", legend = beta.seq, col = 1:length(beta.seq), lwd =
        2, cex = .8)
13  par(new = FALSE)
```

On the other hand, if we change the variable Y as $Y \rightarrow 2Y - 1$ such that Y takes
values in $\{-1, 1\}$, then the distribution can be expressed by

$$P(Y = y \mid x) = \frac{1}{1 + \exp\{-y(\beta_0 + x\beta)\}} . \tag{2.4}$$

In particular, if the realizations of $x \in \mathbb{R}^p$ and $Y \in \{-1, 1\}$ in (2.2) are (x_i, y_i), $i = 1, \ldots, N$, then the likelihood is given by $\prod_{i=1}^{N} \dfrac{1}{1 + e^{-y_i(\beta_0 + x_i\beta)}}$, which is the product
of $P(Y = 1 \mid x_i) = \frac{e^{\beta_0 + x_i\beta}}{1 + e^{\beta_0 + x_i\beta}} = \frac{1}{1 + e^{-(\beta_0 + x_i\beta)}}$ and $\ P(Y = -1 \mid x_i) = \frac{1}{1 + e^{\beta_0 + x_i\beta}}$ for $i = 1, \ldots, N$.

For simplicity, in the following, we identify the maximization of the likelihood
and minimization of the minus log-likelihood L.

$$L := \frac{1}{N} \sum_{i=1}^{N} \log(1 + \exp\{-y_i(\beta_0 + x_i\beta)\}) . \tag{2.5}$$

Let $v_i := \exp\{-y_i(\beta_0 + x_i\beta)\}$ for $i = 1, \ldots, N$, and

$$u; = \begin{bmatrix} \dfrac{y_1 v_1}{1 + v_1} \\ \vdots \\ \dfrac{y_N v_N}{1 + v_N} \end{bmatrix}.$$

Then, the vector $\nabla L \in \mathbb{R}^{p+1}$ whose j-th element is $\dfrac{\partial L}{\partial \beta_j}$, $j = 0, 1, \ldots, p$) can be

expressed by $\nabla L = -\dfrac{1}{N} X^T u$. In fact, we observe $\dfrac{\partial L}{\partial \beta_j} = -\dfrac{1}{N} \sum_{i=1}^{N} x_{i,j} y_i \dfrac{v_i}{1 + v_i}$ ($j =$

$0, 1, \ldots, p$). If we let

$$W = \begin{bmatrix} \dfrac{v_1}{(1 + v_1)^2} & \cdots & 0 \\ \vdots & \ddots & \vdots \\ 0 & \cdots & \dfrac{v_N}{(1 + v_N)^2} \end{bmatrix} \qquad \text{symmetric matrix}$$

then the matrix $\nabla^2 L$ such that the (j, k)-th element is $\dfrac{\partial^2 L}{\partial \beta_j \beta_k}$, $j, k = 0, 1, \ldots, p$ can

be expressed by $\nabla^2 L = \dfrac{1}{N} X^T W X$. In fact, we have

$$\frac{\partial^2 L}{\partial \beta_j \partial \beta_k} = -\frac{1}{N} \sum_{i=1}^{N} x_{i,j} y_i \frac{\partial v_i}{\partial \beta_k} \frac{1}{1 + v_i} = -\frac{1}{N} \sum_{i=1}^{N} x_{i,j} y_i (-y_i x_{i,k}) \frac{v_i}{(1 + v_i)^2}$$

$$= \frac{1}{N} \sum_{i=1}^{N} x_{i,j} x_{i,k} \frac{v_i}{(1 + v_i)^2} \quad (j, k = 0, 1, \ldots, p),$$

where $y_i^2 = 1$.

Next, to obtain the coefficients of logistic regression that maximize the likelihood, we give the initial value of (β_0, β) and repeatedly apply the updates of the Newton method

$$\begin{bmatrix} \beta_0 \\ \beta \end{bmatrix} \leftarrow \begin{bmatrix} \beta_0 \\ \beta \end{bmatrix} - \{\nabla^2 L(\beta_0, \beta)\}^{-1} \nabla L(\beta_0, \beta) \tag{2.6}$$

to obtain the solution $\nabla L(\beta_0, \beta) = 0$.

If we let $z = X \begin{bmatrix} \beta_0 \\ \beta \end{bmatrix} + W^{-1} u$, then we have

$$\begin{bmatrix} \beta_0 \\ \beta \end{bmatrix} - \{\nabla^2 L(\beta_0, \beta)\}^{-1} \nabla L(\beta_0, \beta) = \begin{bmatrix} \beta_0 \\ \beta \end{bmatrix} + (X^T W X)^{-1} X^T u$$

$$= (X^T W X)^{-1} X^T W (X \begin{bmatrix} \beta_0 \\ \beta \end{bmatrix} + W^{-1} u) = (X^T W X)^{-1} X^T W z , \qquad (2.7)$$

which means that we need only to repeatedly execute the steps

1. obtain W, z from β_0, β and
2. obtain β_0, β from W, z.

Example 12 We execute the maximum likelihood solution following the above procedure.

```
1  ## Data Generation
2  N = 1000; p = 2; X = matrix(rnorm(N * p), ncol = p); X = cbind(rep(1,
      N), X)
3  beta = rnorm(p + 1); y = array(N); s = as.vector(X %*% beta); prob = 1
       / (1 + exp(s))
4  for (i in 1:N) {if (runif(1) > prob[i]) y[i] = 1 else y[i] = -1}
5  beta
```

```
1  [1] -0.5859092   0.1610445   0.4134176
```

```
1  ## Computation of the ML solution
2  beta = Inf; gamma = rnorm(p + 1)
3  while (sum((beta - gamma) ^ 2) > 0.001) {
4    beta = gamma
5    s = as.vector(X %*% beta)
6    v = exp(-s * y)
7    u = y * v / (1 + v)
8    w = v / (1 + v) ^ 2
9    z = s + u / w
10   W = diag(w)
11   gamma = as.vector(solve(t(X) %*% W %*% X) %*% t(X) %*% W %*% z)
              ##
12   print(gamma)
13 }
14 beta    ## The true value that we wish to estimate
```

```
1  [1] -0.68982062   0.06228453   0.37459366
```

Repeating the cycle of Newton method, the tentative estimates approach to the true value.

```
1  [1] -0.8248500 -0.3305656 -0.6027963
2  [1] -0.4401921   0.2054357   0.6739074
3  [1] -0.68982062   0.06228453   0.37459366
4  [1] -0.68894021   0.07094424   0.40108031
```

Although the maximum likelihood method can estimate the values of β_0, β, when p is large relative to N, the absolute values of the estimates $\hat{\beta}_0, \hat{\beta}$ may go to infinity. Worse, the larger the value of p, the more likely the procedure diverges. For example, suppose that the rank of X is N and that $N < p$. Then, for any $\alpha_1, \ldots, \alpha_N > 0$, there exists $(\beta_0, \beta) \in \mathbb{R}^{p+1}$ such that $y_i(\beta_0 + x_i\beta) = \alpha_i > 0$ $(i = 1, \ldots, N)$. If we multiply β_0, β by two, the likelihood strictly increases. In general, the larger the value of p is, the more often this phenomenon occurs.

This section considers applying Lasso to obtain a reasonable solution, although it does not maximize the likelihood. The main idea is to regard the almost zero coefficients as zeros to choose relevant variables when p is large. To this end, we add a regularization term to (2.5):

$$L := \frac{1}{N} \sum_{i=1}^{N} \log(1 + \exp\{-y_i(\beta_0 + x_i\beta)\}) + \lambda\|\beta\|_1 \qquad (2.8)$$

and extend the Newton method as follows. We note that (2.7) is the β that minimizes

$$\frac{1}{2N}(z - X\beta)^T W (z - X\beta) . \qquad (2.9)$$

In fact, we observe

$$\nabla \left\{ \frac{1}{2}\left(z - X\begin{bmatrix} \beta_0 \\ \beta \end{bmatrix}\right)^T W \left(z - X\begin{bmatrix} \beta_0 \\ \beta \end{bmatrix}\right)\right\} = X^T W X \begin{bmatrix} \beta_0 \\ \beta \end{bmatrix} - X^T W z .$$

After centralizing X, y introduced in Sect. 2.1, we minimize

$$\frac{1}{2N}(z - X\beta)^T W (z - X\beta) + \lambda\|\beta\|_1 , \qquad (2.10)$$

where the M in Sect. 2.1 is a diagonal matrix with diagonal elements $\sqrt{w_i}$, $i = 1, \ldots, N$. In other words, we need only to repeatedly execute the steps

1. obtain W, z from β_0, β and
2. obtain β_0, β that minimize (2.10) from W, z

(proxy Newton method). Using the function W.linear.lasso in Sect. 2.1, we need to update only ## in the program of Example 12 as follows, where we assume that the leftmost column of X consists of N ones and X contains $p + 1$ columns.

```
1  logistic.lasso = function(X, y, lambda) {
2    p = ncol(X)
3    beta = Inf; gamma = rnorm(p)
4    while (sum((beta - gamma) ^ 2) > 0.01) {
5      beta = gamma
6      s = as.vector(X %*% beta)
7      v = as.vector(exp(-s * y))
```

```
8      u = y * v / (1 + v)
9      w = v / (1 + v) ^ 2
10     z = s + u / w
11     W = diag(w)
12     gamma = W.linear.lasso(X[, 2:p], z, W, lambda = lambda)
13     print(gamma)
14   }
15   return(gamma)
16 }
```

Example 13 Generating data, we examine the behavior of the function `logistic.lasso`.

```
1 N = 100; p = 2; X = matrix(rnorm(N * p), ncol = p); X = cbind(rep(1, N
    ), X)
2 beta = rnorm(p + 1); y = array(N); s = as.vector(X %*% beta); prob = 1
    / (1 + exp(s))
3 for (i in 1:N) {if (runif(1) > prob[i]) y[i] = 1 else y[i] = -1}
4 logistic.lasso(X, y, 0)
```

```
1 [1] -0.4066920  1.0055186  0.7396638
```

```
1 logistic.lasso(X, y, 0.1)
```

```
1 [1] -0.3759565  0.6798822  0.4313101
```

```
1 logistic.lasso(X, y, 0.2)
```

```
1 [1] -0.3489496  0.4094120  0.1710077
```

Example 14 After estimating parameters via logistic regression, we classify each new data point that follows the same distribution. We evaluate the exponent based on the estimated β, whether or not the exponent X is positive (2.2). %*% beta.est is positive or not (2.2).

```
1  ## Data Generation
2  N = 100; p = 2; X = matrix(rnorm(N * p), ncol = p); X = cbind(rep(1, N
     ), X)
3  beta = 10 * rnorm(p + 1); y = array(N); s = as.vector(X %*% beta);
     prob = 1 / (1 + exp(s))
4  for (i in 1:N) {if (runif(1) > prob[i]) y[i] = 1 else y[i] = -1}
5  ## Parameter Estimation
6  beta.est = logistic.lasso(X, y, 0.1)
7  ## Classification
8  for (i in 1:N) {if (runif(1) > prob[i]) y[i] = 1 else y[i] = -1}
9  z = sign(X %*% beta.est)   ## If the exponent is positive, then z=+1,
     otherwise z=-1
10 table(y, z)
```

```
1
2  y     -1   1
3  -1  70    3
4   1   7  20
```

Note that `y` and `z` are the correct answer and the predictive value, respectively. In this example, we predict the correct answer with a precision of 90 %.

On the other hand, the function `glmnet` used for linear regression can be used for logistic regression [11].

Example 15 The following program analyzes the dataset `breastcancer`, which consists of 250 samples (58 cases and 192 control expression data points) w.r.t. 1,001 variables: 1,000 genes (X: covariates) and a case/control (Y: response) of breast cancer.

```
1   library(glmnet)
2   df = read.csv("breastcancer.csv")s
3   ## Put breastcancer.csv in the current directory
4   x = as.matrix(df[, 1:1000])
5   y = as.vector(df[, 1001])
6   cv = cv.glmnet(x, y, family = "binomial")
7   cv2 = cv.glmnet(x, y, family = "binomial", type.measure = "class")
8   par(mfrow = c(1, 2))
9   plot(cv)
10  plot(cv2)
11  par(mfrow = c(1, 1))
```

We evaluate the CV values for each λ and connect the plots (Fig. 2.2). The option `cv.glmnet` (default) evaluates the CV based on the binary deviation

$$-2 \sum_{i:y_i=1} \log(1 + \exp\{-(\hat{\beta}_0 + x_i\hat{\beta})\}) - 2 \sum_{i:y_i=-1} \log(1 + \exp\{\hat{\beta}_0 + x_i\hat{\beta}\})$$

[11], where $\hat{\beta}$ is the estimate from the training data, $(x_1, y_1), \ldots, (x_m, y_m)$ are the test data, and we evaluate the CV by switching the training and test data several times. If we specify the option `type.measure = "class"`, we evaluate the CV based on the error probability.

Using `glmnet`, we obtain the set of relevant genes with nonzero coefficients for the λ that gives the best performance. It seems that the appropriate $\log \lambda$ is between -4 and -3. Setting $\lambda = 0.03$, we execute the following procedure, in which we have used `beta = drop(glm$beta)` rather than the matrix expression `beta = glm$beta` because the former is easier to realize `beta[beta != 0]`.

```
1   glm = glmnet(x, y, lambda = 0.03, family = "binomial")
2   beta = drop(glm$beta); beta[beta != 0]
```

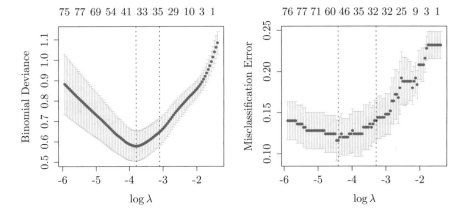

Fig. 2.2 The execution result of Example 15. We evaluated the dataset Breast Cancer for each λ; the left and right figures show the CV evaluations based on the binary deviation and error probability. `glmnet` shows the confidence interval by the upper and lower bounds in the graph [11]. We observe that $\log \lambda$ is the best between -4 and -3. Because we evaluate the CV based on the samples, we show the range of the best λ

2.3 Logistic Regression for Multiple Values

When the covariates take $K \geq 2$ values rather than two, the probability associated with the logistic curves is generalized as

$$P(Y = k \mid x) = \frac{e^{\beta_{0,k} + x\beta^{(k)}}}{\sum_{l=1}^{K} e^{\beta_{0,l} + x\beta^{(l)}}} \quad (k = 1, \dots, K) . \tag{2.11}$$

Thus far, we have regarded $\beta \in \mathbb{R}^p$ as a column vector. In this section, we regard $\beta \in \mathbb{R}^{p \times K}$ as a matrix, $\beta_j = [\beta_{j,1}, \dots, \beta_{j,K}] \in \mathbb{R}^K$ as a row vector, and $\beta^{(k)} = [\beta_{1,k}, \dots, \beta_{p,k}] \in \mathbb{R}^p$ as a column vector. Given the observation $(x_1, y_1), \dots, (x_N, y_N) \in \mathbb{R}^p \times \{1, \dots, K\}$ we can evaluate the negated log-likelihood

$$L := -\frac{1}{N} \sum_{i=1}^{N} \sum_{h=1}^{K} I(y_i = h) \log \frac{\exp\{\beta_{0,h} + x_i \beta^{(h)}\}}{\sum_{l=1}^{K} \exp\{\beta_{0,l} + x_i \beta^{(l)}\}} .$$

Similar to binary logistic regression, we calculate ∇L and $\nabla^2 L$.

Proposition 1 *The differential of L is given by the column vector*

$$\frac{\partial L}{\partial \beta_{j,k}} = -\frac{1}{N} \sum_{i=1}^{N} x_{i,j} \{I(y_i = k) - \pi_{k,i}\} ,$$

where

$$\pi_{k,i} := \frac{\exp(\beta_{0,k} + x_i \beta^{(k)})}{\sum_{l=1}^{K} \exp(\beta_{0,l} + x_i \beta^{(l)})}.$$

For the proof, see the appendix.

Proposition 2 *The twice differential of L is given by the matrix*

$$\frac{\partial^2 L}{\partial \beta_{j,k} \beta_{j',k'}} = \frac{1}{N} \sum_{i=1}^{N} x_{i,j} x_{i,j'} w_{i,k,k'},$$

where $w_{i,k,k'}$ is defined as

$$w_{i,k,k'} := \begin{cases} \pi_{i,k}(1 - \pi_{i,k}), & k' = k \\ -\pi_{i,k} \pi_{i,k'}, & k' \neq k \end{cases}. \tag{2.12}$$

For the proof, see the appendix.

Proposition 3 (Gershgorin) *Let $A = (a_{i,j}) \in \mathbb{R}^{n \times n}$ be a symmetric matrix. If $a_{i,i} \geq \sum_{j \neq i} |a_{i,j}|$ for all $i = 1, \ldots, n$, then A is nonnegative definite.*

For the proof, see the appendix.

Proposition 4 *L is convex.*

Proof $W_i = (w_{i,k,k'}) \in \mathbb{R}^{K \times K}$ is nonnegative definite. In fact, from Proposition 2, we have

$$w_{i,k,k} = \pi_{i,k}(1 - \pi_{i,k}) = \pi_{i,k} \sum_{k' \neq k} \pi_{i,k'} = \sum_{k' \neq k} |w_{i,k,k'}|$$

for all $k = 1, \ldots, K$, and the condition of Proposition 3 is satisfied. Therefore, for any $\gamma = (\gamma_{j,k}) \in \mathbb{R}^{p \times K}$, we have

$$\sum_{j=1}^{p} \sum_{k=1}^{K} \sum_{j'=1}^{p} \sum_{k'=1}^{K} \gamma_{j,k} \frac{\partial^2 L}{\partial \beta_{j,k} \partial \beta_{j',k'}} \gamma_{j'k'}$$

$$= \sum_{j=1}^{p} \sum_{k=1}^{K} \sum_{j'=1}^{p} \sum_{k'=1}^{K} \gamma_{j,k} \left\{ \frac{1}{N} \sum_{i=1}^{N} x_{i,j} w_{i,k,k'} x_{i,j'} \right\} \gamma_{j'k'}$$

$$= \frac{1}{N} \sum_{i=1}^{N} \sum_{k=1}^{K} \sum_{k'=1}^{K} (\sum_{j=1}^{p} x_{i,j} \gamma_{j,k}) w_{i,k,k'} (\sum_{j'=1}^{p} x_{i,j'} \gamma_{j',k'}) \geq 0,$$

and L is convex. □

From a similar discussion as in the previous section, we have the procedure that minimizes

$$L := -\frac{1}{N} \sum_{i=1}^{N} \sum_{h=1}^{K} I(y_i = h) \log \frac{\exp\{\beta_{0,h} + x_i \beta^{(h)}\}}{\sum_{l=1}^{K} \exp\{\beta_{0,l} + x_i \beta^{(l)}\}} + \lambda \sum_{k=1}^{K} \sum_{j=1}^{p} \|\beta_{j,k}\| \quad (2.13)$$

The second term in (2.8) is extended as

$$\lambda \sum_{k=1}^{K} \|\beta_k\|_1 = \lambda \sum_{k=1}^{K} \sum_{j=1}^{p} |\beta_{j,k}| .$$

Because the computation is rather complicated, we often approximate the Taylor expansion such that the nondiagonal elements of $\nabla^2 L$ are zero for $k' \neq k$. Because the objective function is convex, there will be no problem with the convergence.

We note that the value of (2.11) does not change even if we subtract $\gamma_0 + x\gamma$ from all the exponents in the numerator and denominator for any $\gamma_0 \in \mathbb{R}$, $\gamma \in \mathbb{R}^p$. Thus, the first term of (2.13) does not change even if we replace $\beta_{0,k} + x\beta^{(k)}$ by $\beta_{0,k} - \gamma_0 + x(\beta^{(k)} - \gamma)$. However, the second term changes. The γ_j that minimizes $\gamma = (\gamma_1, \ldots, \gamma_p)$, $\sum_{k=1}^{K} |\beta_{j,k} - \gamma_j|$ is the median of $\beta_{j,1}, \ldots, \beta_{j,K}$.

Proposition 5 *For an ordered sequence $a_1 \leq \cdots \leq a_n$ of finite length, $x = a_{m+1}$ and $x = (a_m + a_{m+1})/2$ minimize $f(x) = \sum_{i=1}^{n} |x - a_i|$ when $n = 2m + 1$ and $n = 2m$, respectively. Thus, the median of $a_1, \ldots, a_n$ minimizes the function f.*

For the proof, see the appendix.

By minimizing (2.13), the condition is satisfied (the median is chosen).

Based on the above discussion, we execute the following procedure, in which the variable v in `logistic.lasso` ranges over $k = 1, \ldots, K$.

```
1  multi.lasso = function(X, y, lambda) {
2    X = as.matrix(X)
3    p = ncol(X)
4    n = nrow(X)
5    K = length(table(y))
6    beta = matrix(1, nrow = K, ncol = p)
7    gamma = matrix(0, nrow = K, ncol = p)
8    while (norm(beta - gamma, "F") > 0.1) {
9      gamma = beta
10     for (k in 1:K) {
11       r = 0
12       for (h in 1:K) {if (k != h) r = r + exp(as.vector(X %*% beta[h,
         ]))}
13       v = exp(as.vector(X %*% beta[k, ])) / r
14       u = as.numeric(y == k) - v / (1 + v)
15       w = v / (1 + v) ^ 2
16       z = as.vector(X %*% beta[k, ]) + u / w
17       beta[k, ] = W.linear.lasso(X[, 2:p], z, diag(w), lambda = lambda
         )
18       print(beta[k, ])
19     }
20     for (j in 1:p) {
21       med = median(beta[, j])
```

```
22        for (h in 1:K) beta[h, j] = beta[h, j] - med
23      }
24    }
25    return(beta)
26 }
```

However, the value of $\beta_{0,k}$ is not unique. In the glmnet package, to maintain unique-ness, the values of $\beta_{0,k}, k = 1, \ldots, K$, are determined such that $\sum_{k=1}^{K} \beta_{0,k} = 0$ [11]. When we obtain $\beta_{0,1}, \ldots, \beta_{0,K}$, we subtract its arithmetic mean from them.

Example 16 *(Fisher's Iris)* Using the function multi.lasso, we applied multi-ple value logistic regression to Fisher's Iris dataset to obtain the maximum likelihood estimates $\hat{\beta} \in \mathbb{R}^{p \times K}$ and the predictions $X\hat{\beta}$. We found that all $\hat{\beta}_{2,k}$ are zeros, which is because for each j, $\beta_{j,k} = 0$ for only one k such that $\beta_{j,k}$ is the median. As a result, for the first 50, the next 50, and the last 50 rows, the "1", "2", and "3" have the largest values of $X\hat{\beta}$, respectively.

```
1 df = iris
2 x = matrix(0, 150, 4); for (j in 1:4) x[, j] = df[[j]]
3 X = cbind(1, x)
4 y = c(rep(1, 50), rep(2, 50), rep(3, 50))
5 beta = multi.lasso(X, y, 0.01)
6 X %*% t(beta)
```

Example 17 *(Fisher's Iris)* For Example 16, we obtained the optimum λ via the two CV procedures using the R package glmnet [11], which we show in Fig. 2.3.

```
1  library(glmnet)
2  df = iris
3  x = as.matrix(df[, 1:4]); y = as.vector(df[, 5])
4  n = length(y); u = array(dim = n)
5  for (i in 1:n) if (y[i] == "setosa") u[i] = 1 else
6     if (y[i] == "versicolor") u[i] = 2 else u[i] = 3
7  u = as.numeric(u)
8  cv = cv.glmnet(x, u, family = "multinomial")
9  cv2 = cv.glmnet(x, u, family = "multinomial", type.measure = "class")
10 par(mfrow = c(1, 2)); plot(cv); plot(cv2); par(mfrow = c(1, 1))
11 lambda = cv$lambda.min; result = glmnet(x, y, lambda = lambda, family
       = "multinomial")
12 beta = result$beta; beta.0 = result$a0
13 v = rep(0, n)
14 for (i in 1:n) {
15   max.value = -Inf
16   for (j in 1:3) {
17     value = beta.0[j] + sum(beta[[j]] * x[i, ])
18     if (value > max.value) {v[i] = j; max.value = value}
19   }
20 }
21 table(u, v)
```

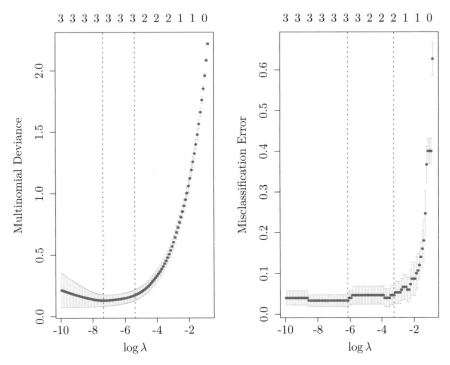

Fig. 2.3 We draw the scores obtained via `glmnet` for each λ, in which the left and right scores are the multiple deviance and error probability, respectively. We may choose the λ that minimizes the score. However, if the sample size is small, some blurring occurs. Thus, the upper and lower bounds that specify the confidence interval are displayed. The figures 0, 1, 2, 3 in the upper locations indicate how many variables are chosen as nonzero for the λ chosen.

```
        z
y      -1   1
      -1  70   3
       1   7  20
```

2.4 Poisson Regression

Suppose that there exists $\mu > 0$ such that the distribution of a random variable Y that ranges over the nonnegative integers is

$$P(Y = k) = \frac{\mu^k}{k!} e^{-\mu} \quad (k = 0, 1, 2, \ldots) \tag{2.14}$$

(Poisson distribution). One can check that μ coincides with the mean value of Y. Hereafter, we assume that μ depends on $x \in \mathbb{R}^p$ and can be expressed by

$$\mu(x) = E[Y \mid X = x] = e^{\beta_0 + x\beta} \quad (x \in \mathbb{R}^p) .$$

Given the observations $(x_1, y_1), \ldots, (x_N, y_N)$, the likelihood is

$$\prod_{i=1}^{N} \frac{\mu_i^{y_i}}{y_i!} e^{-\mu_i} \tag{2.15}$$

with $\mu_i := \mu(x_i) = e^{\beta_0 + x_i\beta}$. We consider applying Lasso when we estimate the parameters β_0, β: minimize the negated log-likelihood

$$L(\beta_0, \beta) := -\frac{1}{N} \sum_{i=1}^{N} \{y_i(\beta_0 + x_i\beta) - e^{\beta_0 + x_i\beta}\}$$

added by the regularization term as

$$L(\beta_0, \beta) + \lambda\|\beta\|_1. \tag{2.16}$$

Similar to logistic regression, if we write $\nabla L = -\dfrac{1}{N}X^T u$ and $\nabla^2 L = X^T W X$, we have

$$u = \begin{bmatrix} y_1 - e^{\beta_0 + x_1\beta} \\ \vdots \\ y_N - e^{\beta_0 + x_N\beta} \end{bmatrix}$$

and

$$W = \begin{bmatrix} e^{\beta_0 + x_1\beta} & \cdots & 0 \\ \vdots & \ddots & \vdots \\ 0 & \cdots & e^{\beta_0 + x_N\beta} \end{bmatrix} .$$

For example, we may construct the procedure as follows:

```
poisson.lasso = function(X, y, lambda) {
  beta = rnorm(p + 1); gamma = rnorm(p + 1)
  while (sum((beta - gamma) ^ 2) > 0.0001) {
    beta = gamma
    s = as.vector(X %*% beta)
    w = exp(s)
    u = y - w
    z = s + u / w
    W = diag(w)
    gamma = W.linear.lasso(X[, 2:(p + 1)], z, W, lambda)
    print(gamma)
  }
```

```
13 │     return(gamma)
14 │ }
```

Example 18 We examined sparse Poisson regression using artificial data.

```
1 │ n = 100; p = 3
2 │ beta = rnorm(p + 1)
3 │ X = matrix(rnorm(n * p), ncol = p); X = cbind(1, X)
4 │ s = as.vector(X %*% beta)
5 │ y = rpois(n, lambda = exp(s))
6 │ beta
7 │ poisson.lasso(X, y, 0.2)
```

The function `poisson.lasso` constructed in this book requires much time to be executed. For business, we might want to use the function `glmnet(X, y, family = "poisson")` [11].

Example 19 For the dataset `birthwt` in the R package `MASS` for the birth rate, we executed Poisson regression. The meanings of the variables are listed in Table 2.1. Because the first variable (whether the birth weight is over 2.5 kg) overlaps with the last variable (the birth weight), we deleted it. We regarded the number of physician visits during the first trimester as the response and executed regression based on the other eight covariates (# samples: $N = 189$), where we chose as the λ value the

Table 2.1 The meanings of the variables in the `birthwt` dataset

Column	Variable name	Meaning
1	low	Indicator of birth weight less than 2.5 kg (0 = higher, 1 = lower)
2	age	Mother's age in years
3	lwt	Mother's weight in pounds at last menstrual period
4	race	Mother's race (1 = white, 2 = black, 3 = other)
5	smoke	Smoking status during pregnancy (0 = no, 1 = yes)
6	ptl	Number of previous premature labors
7	ht	History of hypertension (0 = no, 1 = yes)
8	ui	Presence of uterine irritability (0 = no, 1 = yes)
9	ftv	Number of physician visits during the first trimester
10	bwt	Birth weight in grams

optimum value based on the CV. As a result, we found that mother's age (age), mother's weight (lwt), and mother's hypertension (ht) are the main factors.

```
1   library(glmnet)
2   library(MASS)
3   data(birthwt)
4   df = birthwt[, -1]
5   dy = df[, 8]
6   dx = data.matrix(df[, -8])
7   cvfit = cv.glmnet(x = dx, y = dy, family = "poisson", standardize =
        TRUE)
8   coef(cvfit, s = "lambda.min")
```

```
9 x 1 sparse Matrix of class "dgCMatrix"
                      1
(Intercept)  -1.180159594
age           0.030900888
lwt           0.001653569
race          .
smoke         .
ptl           .
ht           -0.007016192
ui            .
bwt
```

2.5 Survival Analysis

In this section, we consider survival analysis. The problem setting is similar to the ones we considered thus far: finding the relation between covariates and the response from N tuples of the response and p covariate values. For survival analysis, we not only assume that the response Y (survival time) takes positive values y but also allow the response to take the form $Y \geq y$. If an individual dies, the exact survival time is obtained. However, if the survey is terminated before the time of death (the individual changes hospitals, etc.), we regard the survival time to have exceeded the time. To distinguish between the two cases, for the latter, we add the symbol +. The reason we consider the latter case is to utilize the sample: even if the survey is terminated before the time of death, some part of the data can be used to estimate the survival analysis.

We estimate the covariate coefficients based on past data and predict the survival time for a future individual. If we apply Lasso to the problem, we can determine which covariates explain survival.

Example 20 From the kidney dataset, we wish to estimate the kidney survival time. The meanings of the covariates are listed in Table 2.2. The symbol status = 0 expresses survey termination before death and means that the response takes

Table 2.2 The meaning of the variables in the `kidney` dataset

Column	Variable	Meaning
1	Id	Patient ID
2	Time	Time
3	Status	0: survey termination, 1: death
4	Age	Age
5	Sex	Sex (male: 1, female: 2)
6	Disease	0: GN, 1: AN, 2: PKD, 3: Other
7	Frail	Frailty estimate

a larger value than `time`. On the other hand, `status` = 1 expresses death and means that the covariate takes the same value of `time`. The rightmost four are the covariates.

```
1  library(survival)
2  data(kidney)
3  kidney
```

```
1     id time status age sex disease frail
2  1   1    8      1  28   1   Other   2.3
3  2   1   16      1  28   1   Other   2.3
4  3   2   23      1  48   2      GN   1.9
5  4   2   13      0  48   2      GN   1.9
6  . . . . . . . . . . . . . . . . . . . . . . . . . . . . . . . . . . . . .
```

```
1  y = kidney$time
2  delta = kidney$status
3  Surv(y, delta)
```

```
1  [1]    8    16    23   13+   22    28   447   318    30    12    24   245     7
2  [14]   9   511    30    53   196   15   154     7   333   141    8+   96     38
3  [27] 149+   70+  536   25+   17     4+  185   177   292   114   22+  159+    15
4  [40] 108+  152   562   402   24+   13    66    39    46+   12    40  113+   201
5  [53] 132   156    34    30    2     25   130    26    27    58    5+   43    152
6  [66]  30   190     5+  119    8     54+   16+    6+   78    63    8+
```

As marked in the last set of data, due to the survey termination, if the survival time is longer, the symbol + follows the time.

Now, we consider the theoretical framework of survival analysis (Cox model). By $P(t < Y < t + \delta \mid Y \geq t)$, we express the probability of an individual surviving time t through $t + \delta$ ($t < Y < t + \delta$) given the event that he survives at time t ($Y \geq t$). By $S(t)$, we denote the probability (survival function) that the survival time T is at least t (the individual lives at time t). In survival analysis, we define the function by

Fig. 2.4 The Kaplan-Meier curve for each of the kidney diseases

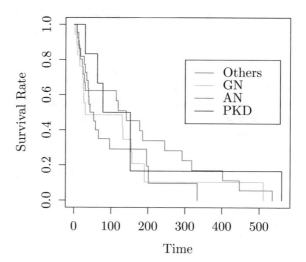

$$h(t) := \lim_{\delta \to 0} \frac{P(t < Y < t + \delta \mid Y \geq t)}{\delta}$$

or, equivalently, by

$$h(t) = -\frac{S'(t)}{S(t)}$$

(hazard function). We express the function h by the product of $h_0(t)$ that does not depend on the covariates $x \in \mathbb{R}^p$ (row vector) and $\exp(x\beta)$:

$$h(t) = h_0(t) \exp(x\beta) ,$$

where the values of $\beta \in \mathbb{R}^p$ are unknown and need to be estimated. In particular, we regard that the intercept β_0 is contained in $h_0(t)$ and assume $\beta_0 = 0$.

Given observations $(x_1, y_1), \ldots, (x_N, y_N) \in \mathbb{R}^p \times \mathbb{R}_{\geq 0}$, we estimate $\beta \in \mathbb{R}^p$ that maximizes the likelihood of $h(t)$, where the function $h_0(t)$ is not used because it does not depend on β. If p is large, i.e., under sparse situations, similar to generalized linear regression, we can apply the L1 regularization.

The main issue of survival analysis is to identify the covariates that determine the survival time.

Example 21 From the `kidney` dataset in the R package `survival`, we depicted the survival rates as time passes in Fig. 2.4 for the diseases AN, GN, and PKD. The graph is stepwise because we made estimations from data. We show how to obtain this graph later.

```
1  fit = survfit(Surv(time, status) ~ disease, data = kidney)
2  plot(fit, xlab = "time", ylab = "survival rate", col = c("red", "green
       ", "blue", "black"))
```

```
3   legend(300, 0.8, legend = c("Other", "GN", "AN", "PKD"),
4          lty = 1, col = c("red", "green", "blue", "black"))
```

Note that the variable `time` in Example 21 takes survival times before death and survey termination. If we remove the times before survey termination, then the number of samples decreases for the estimation. However, if we follow the estimation procedure below, we can utilize the data before survey termination.

If there is no survey termination, then we estimate $S(t)$

$$\hat{S}(t) := \frac{1}{N} \sum_{i=1}^{N} \delta(t_i > t) \tag{2.17}$$

from N death times $t_1 \le t_2 \le \cdots \le t_{N-1} \le t_N$, where $\delta(t_i > t)$ is one for $t_i > t$ and zero otherwise. If $t_1 < \cdots < t_N$, the function is stepwise and decreases by $1/N$ at each $t = t_i$.

Next, we consider the general case in which a survey termination can occur. Let $d_i \ge 1$ $(i = 1, \ldots, k)$ be the number of deaths at time $[t_i, t_{i+1})$, and assume that $t_1 < \cdots < t_k$. Then, the number of total deaths is $D := \sum_{i=1}^{k} d_i \le N$, and $N - D$ survey terminations occur if we assume that the total number of samples is N. If we denote by m_j the number of survey terminations in the interval $[t_j, t_{j+1})$, then the number n_j of surviving individuals immediately before time t_j can be expressed by

$$n_j := \sum_{i=j}^{k} (d_i + m_i)$$

.

We define the Kaplan-Meier estimate as

$$\hat{S}(t) = \begin{cases} 1, & t < t_1 \\ \prod_{i=1}^{l} \frac{n_i - d_i}{n_i}, & t \ge t_1 \end{cases} \tag{2.18}$$

for $t_l \le t < t_{l+1}$, $l = 1, \ldots, k$. Since $n_k = d_k + m_k$, if $m_k = 0$, then $\hat{S}(t) = 0$ for $t > t_k$. Otherwise, if $m_k > 0$, then $\hat{S}(t) > 0$ for $t > t_k$.

If there is no survey termination, then $n_j - d_j = n_{j+1}$ from $n_j = \sum_{i=j}^{k} d_i$, which means that (2.18) becomes

$$\hat{S}(t) = \frac{n_2}{n_1} \times \frac{n_3}{n_2} \times \cdots \times \frac{n_{l+1}}{n_l} = \frac{n_{l+1}}{n_1} = \frac{n_{l+1}}{N}$$

and coincides with (2.17).

Cox (1972) proposed maximizing the partial likelihood

$$\prod_{i:\delta_i=1} \frac{e^{x_i\beta}}{\sum_{j\in R_i} e^{x_j\beta}} \tag{2.19}$$

to estimate the parameter β, where $x_i \in \mathbb{R}^p$ and y_i are covariates and time, respectively, and $\delta_i = 1$ and $\delta_i = 0$ express death and survey termination, respectively, for $i = 1, \ldots, N$.

We define the risk set R_i to be the set of j such that $y_j \geq y_i$ and formulate the Lasso as follows [25]:

$$-\frac{1}{N} \sum_{i:\delta_i=1} \log \frac{e^{x_i\beta}}{\sum_{j\in R_i} e^{x_j\beta}} + \lambda\|\beta\|_1 . \tag{2.20}$$

To obtain the solution, we compute u, W similarly to logistic regression and Poisson regression. In the following, we define

$$L := -\sum_{i:\delta_i=1} \log \frac{e^{x_i\beta}}{\sum_{j\in R_i} e^{x_j\beta}} ,$$

and let $\delta_i = 1, j \in R_i \Longleftrightarrow i \in C_j$.

Proposition 6 *The once and twice differentials of L are*

$$\frac{\partial L}{\partial \beta_k} = -\sum_{i=1}^{N} x_{i,k} \left\{ \delta_i - \sum_{j\in C_i} \frac{e^{x_i\beta}}{\sum_{h\in R_j} e^{x_h\beta}} \right\}$$

$$\frac{\partial^2 L}{\partial \beta_k \partial \beta_l} = \sum_{i=1}^{N} \sum_{h=1}^{N} x_{i,k} x_{h,l} \sum_{j\in C_i} \frac{e^{x_i\beta}}{(\sum_{r\in R_j} e^{x_r\beta})^2} \{I(i=h) \sum_{s\in R_j} e^{x_s\beta} - I(h \in R_j)e^{x_h\beta}\} .$$

In particular, L is convex.

For the proof, see the appendix.

Hence, $u = (u_i) \in \mathbb{R}^N$ and $W = (w_{i,h}) \in \mathbb{R}^{N\times N}$ such that $\dfrac{\partial L}{\partial \beta_k} = -X^T u$ and $\dfrac{\partial^2 L}{\partial \beta_k \partial \beta_l} = X^T W X$ are given as follows:

$$u_i := \delta_i - \sum_{j\in C_i} \frac{e^{x_i\beta}}{\sum_{h\in R_j} e^{x_h\beta}}$$

$$w_{i,h} := \sum_{j\in C_i} \frac{e^{x_i\beta}}{(\sum_{r\in R_j} e^{x_r\beta})^2} \{I(i=h) \sum_{s\in R_j} e^{x_s\beta} - I(h \in R_j)e^{x_h\beta}\} ,$$

where $x_1, \ldots, x_N \in \mathbb{R}^p$ are the rows of $X \in \mathbb{R}^{N \times p}$. In particular, $j \in C_i$ means $h \in R_j$ when $i = h$, and if we let

$$\pi_{i,j} := \frac{e^{x_i \beta}}{\sum_{h \in R_j} e^{x_h \beta}} \, ,$$

then the diagonal elements of W are

$$w_i := \sum_{j \in C_i} \frac{e^{x_i \beta}}{\sum_{h \in R_j} e^{x_h \beta}} \left(1 - \frac{e^{x_i \beta}}{\sum_{h \in R_j} e^{x_h \beta}} \right) = \sum_{j \in C_i} \pi_{i,j} (1 - \pi_{i,j}) \, .$$

Moreover, we can write $u_i = \delta_i - \sum_{j \in C_i} \pi_{i,j}$.

Based on the above discussion, we construct the function cox.lasso below. To avoid a complicated computation, we approximate the off-diagonal elements of W to be zero. Because the objective function is convex, there will be no problem with convergence.

```
cox.lasso = function(X, y, delta, lambda = lambda) {
  delta[1] = 1
  n = length(y)
  w = array(dim = n); u = array(dim = n)
  pi = array(dim = c(n, n))
  beta = rnorm(p); gamma = rep(0, p)
  while (sum((beta - gamma) ^ 2) > 10 ^ {-4}) {
    beta = gamma
    s = as.vector(X %*% beta)
    v = exp(s)
    for (i in 1:n) {for (j in 1:n) pi[i, j] = v[i] / sum(v[j:n])}
    for (i in 1:n) {
      u[i] = delta[i]
      w[i] = 0
      for (j in 1:i) if (delta[j] == 1) {
        u[i] = u[i] - pi[i, j]
        w[i] = w[i] + pi[i, j] * (1 - pi[i, j])
      }
    }
    z = s + u / w; W = diag(w)
    print(gamma)
    gamma = W.linear.lasso(X, z, W, lambda = lambda)[-1]
  }
  return(gamma)
}
```

Example 22 We apply the dataset kidney to the function cox.lasso. The procedure converges for all values of λ, and the estimates coincide with the ones computed via glmnet[11].

```
df = kidney
index = order(df$time)
df = df[index, ]
```

```
4   n = nrow(df); p = 4
5   y = as.numeric(df[[2]])
6   delta = as.numeric(df[[3]])
7   X = as.numeric(df[[4]])
8   for (j in 5:7) X = cbind(X, as.numeric(df[[j]]))
9   z = Surv(y, delta)
10  cox.lasso(X, y, delta, 0)
```

```
[1] 0 0 0 0
[1]   0.0101287 -1.7747758 -0.3887608   1.3532378
[1]   0.01462571 -1.69299527 -0.41598742   1.38980788
[1]   0.01591941 -1.66769665 -0.42331475   1.40330234
[1]   0.01628935 -1.66060178 -0.42528537   1.40862969
```

```
1   cox.lasso(X, y, delta, 0.1)
```

```
[1] 0 0 0 0 [1]   0.00000000 -1.04944510 -0.08990115   1.00822550 [1]
0.00000000 -0.98175893 -0.06107446   0.97534148 [1]   0.00000000
-0.96078476 -0.05449001   0.96180929 [1]   0.00000000 -0.95475614
-0.05306296   0.95673222
```

```
1   cox.lasso(X, y, delta, 0.2)
```

```
[1] 0 0 0 0
[1]   0.0000000 -0.5366227   0.0000000   0.7234343
[1]   0.0000000 -0.5142360   0.0000000   0.6890634
[1]   0.0000000 -0.5099687   0.0000000   0.6800883
```

```
1   glmnet(X, z, family = "cox", lambda = 0.1)$beta
```

```
4 x 1 sparse Matrix of class "dgCMatrix"
            s0
X   .
  -0.87359015
  -0.05659599
   0.92923820
```

Example 23 We downloaded the survival time data of patients with lymphoma, i.e., 1846-2568-2-SP.rda (Alizadeh, 2000) [1] https://www.jstatsoft.org/rt/suppFileMetadata/v039i05 /0/722, and executed the following procedure:

```
1   library(survival)
2   load("LymphomaData.rda"); attach("LymphomaData.rda")
3   names(patient.data); x = t(patient.data$x)
4   y = patient.data$time; delta = patient.data$status; Surv(y, delta)
```

Fig. 2.5 To examine how much the 28 genes with nonzero coefficients affect the survival time, we drew the Kaplan-Meier curves for the sample sets with either positive or negative $z := X\hat{\beta}$ to find a significant difference in the survival times between them

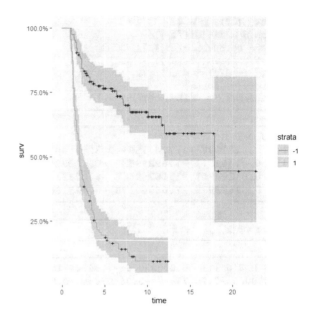

The variables x, y, and delta store the expression data of $p = 7,399$ genes, the times, and the death $(= 1)$/survey termination $(= 0)$ for $N = 240$ samples. We executed the following codes to obtain the λ that minimizes the CV. We found that only 28 coefficients of genes out of 7,399 are nonzero and concluded that the other genes do not affect the survival time.

Moreover, for the $\hat{\beta}$ in which only 28 elements are nonzero, we computed $z = X\hat{\beta}$. We decided that those variables are likely to determine the survival times and drew the Kaplan-Meier curves for the samples with $z_i > 0\,(1)$ and $z_i < 0\,(-1)$ to compare the survival times (Fig. 2.5).

```
1  library(ranger); library(ggplot2); library(dplyr); library(ggfortify)
2  cv.fit = cv.glmnet(x, Surv(y, delta), family = "cox")
3  fit2 = glmnet(x, Surv(y, delta), lambda = cv.fit$lambda.min, family =
       "cox")
4  z = sign(drop(x %*% fit2$beta))
5  fit3 = survfit(Surv(y, delta) ~ z)
6  autoplot(fit3)
7  mean(y[z == 1])
8  mean(y[z == -1])
```

Appendix: Proof of Propositions

Proposition 1 *The differential of L is given by the column vector*

$$\frac{\partial L}{\partial \beta_{j,k}} = -\frac{1}{N}\sum_{i=1}^{N} x_{i,j}\{I(y_i = k) - \pi_{k,i}\},$$

where

$$\pi_{k,i} := \frac{\exp(\beta_{0,k} + x_i \beta^{(k)})}{\sum_{l=1}^{K}\exp(\beta_{0,l} + x_i \beta^{(l)})}.$$

Proof Differentiating

$$L := -\frac{1}{N}\sum_{i=1}^{N}\sum_{h=1}^{K} I(y_i = h)\log\frac{\exp\{\beta_{0,h} + x_i\beta_h\}}{\sum_{l=1}^{K}\exp\{\beta_{0,l} + x_i\beta_l\}}$$

$$= -\frac{1}{N}\sum_{i=1}^{N}\sum_{h=1}^{K} I(y_i = h)$$

$$\left\{(\beta_{0,h} + x_i\beta_h) - \log\left[\sum_{l=1}^{K}\exp(\beta_{0,l} + x_i\beta_l)\right]\right\}$$

by $\beta_{j,k}$ $(j = 1, \ldots, p,\ k = 1, \ldots, K)$, we obtain

$$\frac{\partial L}{\partial \beta_{j,k}} = -\frac{1}{N}\sum_{i=1}^{N} x_{i,j}\sum_{h=1}^{K} I(y_i = h)\left\{I(h = k) - \frac{\exp\{\beta_{0,k} + x_i\beta_k\}}{\sum_{l=1}^{K}\exp\{\beta_{0,l} + x_i\beta_l\}}\right\}$$

$$= -\frac{1}{N}\sum_{i=1}^{N} x_{i,j}\left\{I(y_i = k) - \frac{\exp\{\beta_{0,k} + x_i\beta_k\}}{\sum_{l=1}^{K}\exp\{\beta_{0,l} + x_i\beta_l\}}\right\}$$

$$= -\frac{1}{N}\sum_{i=1}^{N} x_{i,j}\{I(y_i = k) - \pi_{k,i}\}.$$

$\square$

Proposition 2 *The twice differential of L is given by the matrix*

$$\frac{\partial^2 L}{\partial \beta_{j,k}\beta_{j',k}} = \frac{1}{N}\sum_{i=1}^{N} x_{i,j}x_{i,j'}w_{i,k,k'}$$

where $w_{i,k,k'}$ is defined as

$$w_{i,k,k'} := \begin{cases} \pi_{i,k}(1 - \pi_{i,k}), & k' = k \\ -\pi_{i,k}\pi_{i,k'}, & k' \neq k \end{cases} . \tag{2.12}$$

Proof For $k' = k$, differentiating by $\beta_{j',k}$ $(j' = 1, \ldots, p)$, we obtain

$$\frac{\partial^2 L}{\partial \beta_{j,k} \partial \beta_{j',k}} k = \frac{\partial}{\partial \beta_{j',k}} \left[-\frac{1}{N} \sum_{i=1}^{N} x_{i,j} \{ I(y_i = k) - \pi_{k,i} \} \right] = \frac{1}{N} \sum_{i=1}^{N} x_{i,j} \frac{\partial \pi_{k,i}}{\partial \beta_{j',k}}$$

$$= \frac{1}{N} \sum_{i=1}^{N} x_{i,j} \frac{x_{i,j'} \exp(\beta_{0,k} + x_i \beta_k) \{\sum_{l=1}^{K} \exp(\beta_{0,l} + x_i \beta_l)\} - x_{i,j'} \{\exp(\beta_{0,k} + x_i \beta_k)\}^2}{\{\sum_{l=1}^{K} \exp(\beta_{0,l} + x_i \beta_l)\}^2}$$

$$= \frac{1}{N} \sum_{i=1}^{N} x_{i,j} x_{i,j'} \pi_{i,k}(1 - \pi_{i,k}) .$$

For $k' \neq k$, differentiating by $\beta_{j',k'}$ $(j' = 1, \ldots, p)$, we obtain

$$\frac{\partial^2 L}{\partial \beta_{j,k} \partial \beta_{j',k'}} = \frac{\partial}{\partial \beta_{j',k'}} \left[-\frac{1}{N} \sum_{i=1}^{N} x_{i,j} \{ I(y_i = k) - \pi_{k,i} \} \right] = \frac{1}{N} \sum_{i=1}^{N} x_{i,j} \frac{\partial \pi_{k,i}}{\partial \beta_{j',k'}}$$

$$= \frac{1}{N} \sum_{i=1}^{N} x_{i,j} \exp(\beta_{0,k} + x_i \beta_k) \left\{ -\frac{x_{i,j'} \exp(\beta_{0,k'} + x_i \beta_{k'})}{\{\sum_{l=1}^{K} \exp(\beta_{0,l} + x_i \beta_l)\}^2} \right\}$$

$$= -\frac{1}{N} \sum_{i=1}^{N} x_{i,j} x_{i,j'} \pi_{i,k} \pi_{i,k'} .$$

Proposition 3 (Gershgorin) *Let $A = (a_{i,j}) \in \mathbb{R}^{n \times n}$ be a symmetric matrix. If $a_{i,i} \geq \sum_{j \neq i} |a_{i,j}|$ for all $i = 1, \ldots, n$, then A is nonnegative definite.*

Proof For any eigenvalue λ of square matrix $A = (a_{j,k}) \in \mathbb{R}^{n \times n}$, there exists an eigenvector $x = [x_1, \ldots, x_n]^T$ and $1 \leq i \leq n$ such that $x_i = 1$ and $|x_j| \leq 1$, $j \neq i$. Thus, from

$$a_{i,i} + \sum_{j \neq i} a_{i,j} x_j = \sum_{j=1}^{n} a_{i,j} x_j = \lambda x_i = \lambda ,$$

we have

$$|\lambda - a_{i,i}| = |\sum_{j \neq i} a_{i,j} x_j| \leq \sum_{j \neq i} |a_{i,j}|$$

and

$$a_{i,i} - \sum_{j \neq i} |a_{i,j}| \leq \lambda \leq a_{i,i} + \sum_{j \neq i} |a_{i,j}| ,$$

at least for one $1 \leq i \leq n$, which means that if $a_{i,i} \geq \sum_{j \neq i} |a_{i,j}|$ for $i = 1, \ldots, n$, then all the eigenvalues are nonnegative, and the matrix A is nonnegative definite. $\square$

Proposition 5 *For an ordered sequence $a_1 \leq \cdots \leq a_n$ of finite length, $x = a_{m+1}$ and $x = (a_m + a_{m+1})/2$ minimize $f(x) = \sum_{i=1}^{n} |x - a_i|$ when $n = 2m + 1$ and $n = 2m$, respectively. Thus, the median of $a_1, \ldots, a_n$ minimizes the function f.*

Proof If $n = 2m + 1$ and $a_{i-1} < a_i = \cdots = a_{m+1} = \cdots = a_j < a_{j+1}, i \leq m + 1 \leq j$, then the subderivative of f at a_{m+1} is

$$(j - i + 1)[-1, 1] - (i - 1) + (n - j) = [n - 2j, n - 2i + 2]$$
$$= [2m - 2j + 1, 2m - 2i + 3] \supseteq [-1, 1]$$

and contains zero. If $n = 2m$ and $a_m \neq a_{m+1}$, then the subderivative of f at $(a_m + a_{m+1})/2$ is zero. Finally, if $n = 2m$ and $a_{i-1} < a_i = \cdots = a_m = a_{m+1} = \cdots = a_j < a_{j+1}, i \leq m, m + 1 \leq j$, then the subderivative of f at $(a_m + a_{m+1})/2$ is

$$(j - i + 1)[-1, 1] - (i - 1) + (n - j) = [n - 2j, n - 2i + 2] = [2m - 2j, 2m - 2i + 2] \supseteq [-2, 2]$$

and contains zero. Thus, for any case, x minimizes $f(x)$. □

Proposition 6 *The once and twice differentials of L are*

$$\frac{\partial L}{\partial \beta_k} = -\sum_{i=1}^{N} x_{i,k} \left\{ \delta_i - \sum_{j \in C_i} \frac{e^{x_i \beta}}{\sum_{h \in R_j} e^{x_h \beta}} \right\}$$

$$\frac{\partial^2 L}{\partial \beta_k \partial \beta_l} = \sum_{i=1}^{N} \sum_{h=1}^{N} x_{i,k} x_{h,l} \sum_{j \in C_i} \frac{e^{x_i \beta}}{(\sum_{r \in R_j} e^{x_r \beta})^2}$$
$$\{I(i = h) \sum_{s \in R_j} e^{x_s \beta} - I(h \in R_j) e^{x_h \beta}\} .$$

In particular, L is convex.

Proof Since for $S_i = \sum_{h \in R_i} e^{x_h \beta}$, we have

$$\sum_{i:\delta_i=1} \sum_{j \in R_i} \frac{x_{j,k} e^{x_j \beta}}{S_i} = \sum_{j=1}^{N} \sum_{i \in C_j} \frac{x_{j,k} e^{x_j \beta}}{S_i} = \sum_{i=1}^{N} x_{i,k} \sum_{j \in C_i} \frac{e^{x_i \beta}}{S_j} ,$$

each element of $\nabla L = -X^T u$ is derived as follows:

$$\frac{\partial L}{\partial \beta_k} = -\sum_{i:\delta_i=1} \left\{ x_{i,k} - \frac{\sum_{j \in R_i} x_{j,k} e^{x_j \beta}}{\sum_{h \in R_i} e^{x_h \beta}} \right\}$$

$$= -\sum_{i=1}^{N} x_{i,k} \left\{ \delta_i - \sum_{j \in C_i} \frac{e^{x_i \beta}}{\sum_{h \in R_j} e^{x_h \beta}} \right\} .$$

On the other hand, each element of $\nabla^2 L = X^T W X$ is derived as follows:

$$
\frac{\partial^2 L}{\partial \beta_k \partial \beta_l} = \sum_{i=1}^{N} x_{i,k} \sum_{j \in C_i} \frac{\partial}{\partial \beta_l} \left(\frac{e^{x_i \beta}}{\sum_{h \in R_j} e^{x_h \beta}} \right)
$$

$$
= \sum_{i=1}^{N} x_{i,k} \sum_{j \in C_i} \frac{1}{(\sum_{r \in R_j} e^{x_r \beta})^2} \{ x_{i,l} e^{x_i \beta} \sum_{s \in R_j} e^{x_s \beta} - e^{x_i \beta} \sum_{h \in R_j} x_{h,l} e^{x_h \beta} \}
$$

$$
= \sum_{i=1}^{N} \sum_{h=1}^{N} x_{i,k} x_{h,l} \sum_{j \in C_i} \frac{e^{x_i \beta}}{(\sum_{r \in R_j} e^{x_r \beta})^2} \{ I(i = h) \sum_{s \in R_j} e^{x_s \beta} - I(h \in R_j) e^{x_h \beta} \} .
$$

Hence, we have

$$
\frac{\partial^2 L}{\partial \beta_k \partial \beta_l} = \sum_{i=1}^{N} \sum_{h=1}^{N} x_{i,k} x_{h,l} w_{i,h} ,
$$

where

$$
w_{i,h} = \begin{cases} \displaystyle\sum_{j \in C_i} \frac{e^{x_i \beta}}{(\sum_{r \in R_j} e^{x_r \beta})^2} \{ \sum_{s \in R_j} e^{x_s \beta} - I(i \in R_j) e^{x_h \beta} \}, & h = i \\ \displaystyle -\sum_{j \in C_i} \frac{e^{x_i \beta}}{(\sum_{r \in R_j} e^{x_r \beta})^2} \{ I(h \in R_j) e^{x_h \beta} \}, & h \neq i \end{cases}.
$$

Moreover, from $i \in R_j$, such $W = (w_{i,j})$ satisfies $w_{i,i} = \sum_{h \neq i} |w_{i,h}|$ for each $i = 1, \ldots, N$. From Proposition 3, W is nonnegative definite. Thus, for arbitrary $[z_1, \ldots, z_p] \in \mathbb{R}^p$, we have

$$
\sum_{k=1}^{p} \sum_{l=1}^{p} \sum_{i=1}^{N} \sum_{h=1}^{N} z_k z_l x_{i,k} x_{h,l} w_{i,h} \geq 0 ,
$$

where L is nonnegative definite (its Hessian is nonnegative definite).   □

Exercises 21–34

21. Let Y be a random variable that takes values in $\{0, 1\}$. We assume that there exist $\beta_0 \in \mathbb{R}$ and $\beta \in \mathbb{R}^p$ such that the conditional probability $P(Y = 1 \mid x)$ given for each $x \in \mathbb{R}^p$ can be expressed by

$$
\log \frac{P(Y = 1 \mid x)}{P(Y = 0 \mid x)} = \beta_0 + x\beta \qquad \text{(cf. (2.2))} \tag{2.21}
$$

(logistic regression).

(a) Prove that (2.21) can be expressed by

$$P(Y = 1 \mid x) = \frac{\exp(\beta_0 + x\beta)}{1 + \exp(\beta_0 + x\beta)} \qquad \text{(cf. (2.3))}. \qquad (2.22)$$

(b) Let $p = 1$ and $\beta_0 = 0$. The following procedure draws the curves by computing the right-hand side of (2.22) for various $\beta \in \mathbb{R}$. Fill in the blanks and execute the procedure for $\beta = 0$. How do the shapes evolve as β grows?

```
1   f = function(x) return(exp(beta.0 + beta * x) / (1 + exp(beta
        .0 + beta * x)))
2   beta.0 = 0; beta.seq = c(0, 0.2, 0.5, 1, 2, 10)
3   m = length(beta.seq)
4   beta = beta.seq[1]
5   plot(f, xlim = c(-10, 10), ylim = c(0, 1), xlab = "x", ylab =
        "y",
6       col = 1, main = "Logistic Curve")
7   for (i in 2:m) {
8     beta = beta.seq[i]
9     par(new = TRUE)
10    plot(f, xlim = c(-10, 10), ylim = c(0, 1), xlab = "", ylab =
        "", axes = FALSE, col = i)
11  }
12  legend("topleft", legend = beta.seq, col = 1:length(beta.seq),
            lwd = 2, cex = .8)
13  par(new = FALSE)
```

(c) For logistic regression (2.21), when the realizations $x \in \mathbb{R}^p$ and $Y \in \{0, 1\}$ are (x_i, y_i) $(i = 1, \ldots, N)$, the likelihood is given by $\displaystyle\prod_{i=1}^{N} \frac{e^{y_i\{\beta_0 + x_i\beta\}}}{1 + e^{\beta_0 + x_i\beta}}$. In its Lasso evaluation

$$-\frac{1}{N} \sum_{i=1}^{N} [y_i(\beta_0 + x_i\beta) - \log(1 + e^{\beta_0 + x_i\beta})] + \lambda\|\beta\|_1, \qquad (2.23)$$

Y takes values in $\{0, 1\}$. However, an alternative expression as in (2.24) is often used, in which Y takes values in $\{-1, 1\}$:

$$\frac{1}{N} \sum_{i=1}^{N} \log(1 + \exp\{-y_i(\beta_0 + x_i\beta)\}) + \lambda\|\beta\|_1 \qquad \text{(cf. (2.8))}. \qquad (2.24)$$

Show that if we replace $y_i = 0$ with $y_i = -1$, then (2.23) is equivalent to (2.24).

Hereafter, we denote by $X \in \mathbb{R}^{N \times (p+1)}$ the matrix such that the (i, j)-th element is $x_{i,j}$ for $j = 1, \ldots, p$ and the leftmost column (the 0th column) is a vector consisting of N ones, and let $x_i = [x_{i,1}, \ldots, x_{i,p}]$. We assume that the random variable Y takes values in $\{-1, 1\}$.

22. For $L(\beta_0, \beta) := \sum_{i=1}^{N} \log\{1 + \exp(-y_i(\beta_0 + x_i\beta))\}$, show the following equations, where $v_i := \exp\{-y_i(\beta_0 + x_i\beta)\}$.

(a) The matrix ∇L such that the j-th element is $\dfrac{\partial L}{\partial \beta_j}$ $(j = 0, 1, \ldots, p)$ can be expressed by $\nabla L = -X^T u$, where

$$u = \begin{bmatrix} \dfrac{y_1 v_1}{1 + v_1} \\ \vdots \\ \dfrac{y_N v_N}{1 + v_N} \end{bmatrix}.$$

(b) The matrix $\nabla^2 L$ such that the (j, k)-th element is $\dfrac{\partial^2 L}{\partial \beta_j \beta_k}$ $(j, k = 0, 1, \ldots, p)$ can be expressed by $\nabla^2 L = X^T W X$, where

$$W = \begin{bmatrix} \dfrac{v_1}{(1 + v_1)^2} & \cdots & 0 \\ \vdots & \ddots & \vdots \\ 0 & \cdots & \dfrac{v_N}{(1 + v_N)^2} \end{bmatrix} \quad \text{diagonal matrix .}$$

Moreover, setting $\lambda = 0$, we construct the procedure that estimates the logistic regression coefficients as follows. First, we give an initial value of (β_0, β) and update the value via the Newton method until it converges. Execute the following, and examine the convergence. In particular, observe that the procedure tends to diverge as we make p larger.

```
1   ## Data Generation
2   N = 1000; p = 2; X = matrix(rnorm(N * p), ncol = p); X = cbind(rep
        (1, N), X)
3   beta = rnorm(p + 1); y = array(N); s = as.vector(X %*% beta); prob
        = 1 / (1 + exp(s))
4   for (i in 1:N) {if (runif(1) > prob[i]) y[i] = 1 else y[i] = -1}
5   beta
6   ## Maximum Likelihood Computation
7   beta = Inf; gamma = rnorm(p + 1)
8   while (sum((beta - gamma) ^ 2) > 0.001) {
9     beta = gamma
10    s = as.vector(X %*% beta)
11    v = exp(-s * y)
12    u = y * v / (1 + v)
13    w = v / (1 + v) ^ 2
14    z = s + u / w
15    W = diag(w)
16    gamma = as.vector(solve(t(X) %*% W %*% X) %*% t(X) %*% W %*% z)
17    print(gamma)
18  }
```

23. Show that the (β_0, β) that maximize the likelihood diverges when $\lambda = 0$.

    (a) $N < p$, and the rank of X is N.
       Hint In Exercise 2.5, if we multiply $-X^T u = 0$ by $-X$ from left to right,
       then it becomes $XX^T u = 0$. Since XX^T is invertible, unless $u = 0$, $\tilde{X}\tilde{X}^T u = 0$ does not reach the stationary solution. Moreover, u cannot be zero for finite
       (β_0, β).

    (b) There exist (β_0, β) such that $y_i(\beta_0 + x_i\beta) > 0$ for all $i = 1, \ldots, N$.
       Hint If such (β_0, β) exist, $(2\beta_0, 2\beta)$ will make L smaller.

24. The following procedure analyzes the dataset `breastcancer`, which consists of 1,000 covariate variables (the expression data of 1,000 genes) and one binary response (case/control) for 256 samples (58 cases and 192 control). Identify the relevant genes that affect breast cancer.

```
1  df = read.csv("breastcancer.csv")
2  x = as.matrix(df[, 1:1000])
3  y = as.vector(df[, 1001])
4  cv = cv.glmnet(x, y, family = "binomial")
5  cv2 = cv.glmnet(x, y, family = "binomial", type.measure = "class")
6  par(mfrow = c(1, 2))
7  plot(cv)
8  plot(cv2)
9  par(mfrow = c(1, 1))
```

The function `cv.glmnet` evaluates the test data based on the binary deviance

$$\frac{1}{N}\sum_{i=1}^{N} -\log P(Y = y_i \mid X = x_i)$$

and error rate if we do not specify anything and if the option is `type.measure = "class"`, respectively.

    Fill in the optimum λ based on the CV, make the following codes, and find the genes with nonzero coefficients.

```
1  glm = glmnet(x, y, lambda = ## Blank ##, family = "binomial")
```

Hint Use `beta = drop(glm$beta)` rather than `beta = glm$beta`. Then, it will be easier to construct a code similar to `beta[beta != 0]`.

25. Logistic regression can be extended as follows if the response takes $K \geq 2$ values rather than binary values.

$$P(Y = k \mid x) = \frac{e^{\beta_{0,k}+x\beta^{(k)}}}{\sum_{l=1}^{K} e^{\beta_{0,l}+x\beta^{(l)}}} \quad (k = 1, \ldots, K). \quad \text{(cf. (2.11))} \quad (2.25)$$

    (a) (2.25) remains correct even if we subtract $\gamma_0 + x\gamma$ from the exponents in the numerator and denominator for any $\gamma_0 \in \mathbb{R}, \gamma \in \mathbb{R}^p$.

    (b) The second term of (2.24) can be extended as $\lambda \sum_{k=1}^{K} \|\beta_k\|_1 = \lambda \sum_{k=1}^{K} \sum_{j=1}^{p} |\beta_{j,k}|$. The first term does not change even if we replace $\beta_{0,k} + x\beta^{(k)}$

by $\beta_{0,k} - \gamma_0 + x(\beta^{(k)} - \gamma)$, but the second term changes. One can check that for $\gamma = (\gamma_1, \ldots, \gamma_p)$, the γ_j value that minimizes $\sum_{k=1}^{K} |\beta_{j,k} - \gamma_j|$ is the median of $\beta_{j,1}, \ldots, \beta_{j,K}$. How can we obtain the minimum Lasso evaluation from the original $\beta_{j,1}, \ldots, \beta_{j,K}$ ($j = 1, \ldots, p$)?

(c) In the `glmnet` package, to maintain the uniqueness of the $\beta_{0,k}$ value, it is set such that $\sum_{k=1}^{K} \beta_{0,k} = 0$. After obtaining the original $\beta_{0,1}, \ldots, \beta_{0,K}$, what computation should be performed to obtain the unique $\beta_{0,1}, \ldots, \beta_{0,K}$?

26. Execute the following procedure for the Iris dataset with $n = 150$ and $p = 4$. Output the two graphs via `cv.glmnet`, and find the optimum λ in the sense of CV. Moreover, find the β_0, β w.r.t. λ, and execute it using the 150 covariates.

```
1  library(glmnet)
2  df = read.table("iris.txt", sep = ",")
3  x = as.matrix(df[, 1:4])
4  y = as.vector(df[, 5])
5  y = as.numeric(y == "Iris-setosa")
6  cv = cv.glmnet(x, y, family = "binomial")
7  cv2 = cv.glmnet(x, y, family = "binomial", type.measure = "class")
8  par(mfrow = c(1, 2))
9  plot(cv)
10 plot(cv2)
11 par(mfrow = c(1, 1))
12 lambda = cv$lambda.min
13 result = glmnet(x, y, lambda = lambda, family = "binomial")
14 beta = result$beta
15 beta.0 = result$a0
16 f = function(x) return(exp(beta.0 + x %*% beta))
17 z = array(dim = 150)
18 for (i in 1:150) z[i] = drop(f(x[i, ]))
19 yy = (z > 1)
20 sum(yy == y)
```

We evaluate the correct rate, not for $K = 2$ but for $K = 3$. Fill in the blanks and execute it.

```
1  library(glmnet)
2  df = read.table("iris.txt", sep = ",")
3  x = as.matrix(df[, 1:4]); y = as.vector(df[, 5]); n = length(y); u
       = array(dim = n)
4  for (i in 1:n) if (y[i] == "Iris-setosa") u[i] = 1 else
5    if (y[i] == "Iris-versicolor") u[i] = 2 else u[i] = 3
6  u = as.numeric(u)
7  cv = cv.glmnet(x, u, family = "multinomial")
8  cv2 = cv.glmnet(x, u, family = "multinomial", type.measure = "
       class")
9  par(mfrow = c(1, 2)); plot(cv); plot(cv2); par(mfrow = c(1, 1))
10 lambda = cv$lambda.min; result = glmnet(x, y, lambda = lambda,
       family = "multinomial")
11 beta = result$beta; beta.0 = result$a0
12 v = array(dim = n)
13 for (i in 1:n) {
14   max.value = -Inf
15   for (j in 1:3) {
```

```
16        value = ## Blank ##
17        if (value > max.value) {v[i] = j; max.value = value}
18     }
19  }
20  sum(u == v)
```

Hint Each `beta` and beta.0 is a list of length 3, and the former is a vector that stores the coefficient values.

27. If the response takes $K \geq 2$ values rather than two, the probability associated with the logistic curve is generalized as follows:

$$P(Y = k \mid x) = \frac{e^{\beta_{0,k}+x\beta^{(k)}}}{\sum_{l=1}^{K} e^{\beta_{0,l}+x\beta^{(l)}}} \quad (k = 1, \ldots, K). \quad \text{(cf. (2.11))}$$

First, given observations $(x_1, y_1), \ldots, (x_N, y_N) \in \mathbb{R}^p \times \{1, \ldots, K\}$, we can compute the negated log-likelihood

$$L := -\frac{1}{N} \sum_{i=1}^{N} \sum_{h=1}^{K} I(y_i = h) \log \frac{\exp\{\beta_{0,h} + x_i \beta^{(h)}\}}{\sum_{l=1}^{K} \exp\{\beta_{0,l} + x_i \beta^{(l)}\}}$$

. Using the fact that the twice differential of L is

$$\frac{\partial^2 L}{\partial \beta_{j,k} \beta_{j',k}} = \begin{cases} \sum_{i=1}^{N} x_{i,j} x_{i,j'} \pi_{i,k}(1 - \pi_{i,k}), & k' = k \\ \sum_{i=1}^{N} x_{i,j} x_{i,j'} \pi_{i,k} \pi_{i,k'}, & k' \neq k \end{cases},$$

we write them as $\sum_{i=1}^{N} x_{i,j} x_{i,j'} w_{i,k,k'}$. Show that the matrix $W_i = (w_{i,k,k'}) \in \mathbb{R}^{K \times K}$ for each $i = 1, 2, \ldots n$ is nonnegative definite.

28. We assume that the parameter $\mu := E[Y] > 0$ of the Poisson distribution

$$P(Y = k) = \frac{\mu^k}{k!} e^{-\mu} \quad (k = 0, 1, 2, \ldots) \quad \text{(cf. (2.14))}$$

can be expressed by

$$\mu(x) = E[Y \mid X = x] = e^{\beta_0 + x\beta} \quad (x \in \mathbb{R}^p)$$

for $x \in \mathbb{R}^p$. Then, given observations $(x_1, y_1), \ldots, (x_N, y_N)$, the likelihood is

$$\prod_{i=1}^{N} \frac{\mu_i^{y_i}}{y_i!} e^{-\mu_i} \quad \text{(cf. (2.15))} \quad (2.26)$$

with $\mu_i := \mu(x_i) = e^{\beta_0 + x_i \beta}$. To obtain β_0, β, we minimize

$$\frac{1}{N} L(\beta_0, \beta) + \lambda \|\beta\|_1 \quad \text{(cf. (2.16))} \quad (2.27)$$

with $L(\beta_0, \beta) := -\sum_{i=1}^{N}\{y_i(\beta_0 + x_i\beta) - e^{\beta_0 + x_i\beta}\}$ if we apply Lasso.

(a) How can we derive (2.27) from (2.26)?

(b) If we write $\nabla L = -X^T u$, show that

$$
u = \begin{bmatrix} y_1 - e^{\beta_0 + x_1\beta} \\ \vdots \\ y_N - e^{\beta_0 + x_N\beta} \end{bmatrix}.
$$

(c) If we write $\nabla L = X^T W X$, show that

$$
W = \begin{bmatrix} e^{\beta_0 + x_1\beta} & \cdots & 0 \\ \vdots & \ddots & \vdots \\ 0 & \cdots & e^{\beta_0 + x_N\beta} \end{bmatrix} \quad \text{(diagonal matrix)} .
$$

We execute Poisson regression for $\lambda \geq 0$. Fill in the blank and execute the procedure.

```
1   ## Data Generation
2   N = 1000
3   p = 7
4   beta = rnorm(p + 1)
5   X = matrix(rnorm(N * p), ncol = p)
6   X = cbind(rep(1, N), X)
7   s = X %*% beta
8   y = rpois(N, lambda = exp(s))
9   beta
10  ## Conputation of the ML estimates
11  lambda = 100
12  beta = Inf
13  gamma = rnorm(p + 1)
14  while (sum((beta - gamma) ^ 2) > 0.01) {
15      beta = gamma
16      s = as.vector(X %*% beta)
17      w = ## Blank (1) ##
18      u = ## Blank (2) ##
19      z = ## Blank (3) ##
20      W = diag(w)
21      gamma = # Blank (4) #
22      print(gamma)
23  }
```

In the following, we consider survival analysis, particularly for the Cox model. Using the random variables $T, C \geq 0$ that express death and survey termination, we define $Y := \min\{T, C\}$. For $t \geq 0$, let $S(t)$ be the probability of the event $T > t$:

$$
h(t) := -\frac{S'(t)}{S(t)}
$$

or equivalently,

$$h(t) := \lim_{\delta \to 0} \frac{P(t < Y < t + \delta \mid Y \geq t)}{\delta}.$$

A Cox model can be expressed by the product of $h_0(t)$ (hazard function) that depends only on the covariate $x \in \mathbb{R}$:

$$h(t) = h_0(t) \exp(x\beta).$$

In particular, we regard the constant β_0 as contained in $h_0(t)$ and assume $\beta_0 = 0$. Given $(x_1, y_1), \ldots, (x_N, y_N) \in \mathbb{R}^p \times \mathbb{R}_{\geq 0}$, we estimate $\beta \in \mathbb{R}^p$ that maximizes the likelihood of $h(t)$. However, h_0 is not used to compute the ML estimate because it is constant. When p is large, i.e., under sparse situations, L1 regularization is considered similar to other regression models.

29. We estimate the survival time of kidney patients from the `kidney` dataset.

Column	Variable	Meaning
1	id	patient ID
2	time	time
3	status	0: survey termination, 1: death
4	age	age
5	sex	sex (male: 1, female: 2)
6	disease	0: GN, 1: AN, 2: PKD, 3: Other
7	frail	frailty estimate

We first execute the following code:

```
library(survival)
data(kidney)
names(kidney)
y = kidney$time
delta = kidney$status
Surv(y, delta)
```

(a) What procedure does the function `Surv` execute?
(b) We draw the survival time curves for each kidney disease. Replace the curves for diseases with those for sex, and add labels and a legend.

```
fit = survfit(Surv(time, status) ~ disease, data = kidney)
plot(fit, xlab = "Time", ylab = "Error Rate", col = c("red", "
    green", "blue", "black"))
legend(300, 0.8, legend = c("Others", "GN", "AN", "PKD"),
        lty = 1, col = c("red", "green", "blue", "black"))
## Execute the below as well.
library(ranger); library(ggplot2); library(dplyr); library(
    ggfortify); autoplot(fit)
```

30. The variable `time` in Exercise 29 takes both the survival time and the time before survey termination into account. Let $t_1 < t_2 < \cdots < t_k$ $(k \leq N)$ be the

ordered survival time (excluding the survey terminations). If there are d_i deaths at time t_i for $i = 1, \ldots, k$, then the total number of deaths is $D = \sum_{j=1}^{k} d_j$. If no survey termination occurs for the N samples, we have $D = N$. If we denote by m_j $(j = 1, \ldots, k)$ the number of survey terminations in the interval $[t_j, t_{j+1})$, then the number (the size of the risk set) of survivors immediately before t_j is

$$n_j = \sum_{i=j}^{k} (d_i + m_i) \quad (j = 1, \ldots, k) .$$

Then, the probability $S(t)$ of the survival time T being larger than t can be estimated as (Kaplan-Meier estimate): for $t_l \le t < t_{l+1}$,

$$\hat{S}(t) = \begin{cases} 1, & t < t_1 \\ \displaystyle\prod_{i=1}^{l} \frac{n_i - d_i}{n_i}, & t \ge t_1. \end{cases} \quad \text{(cf. (2.18))}$$

What do the estimates become when no survey termination occurs during $t_l \le t < t_{l+1}, l = 1, \ldots, k$?

31. Cox (1972) proposed maximizing the partial likelihood function

$$\prod_{i:\delta_i=1} \frac{e^{x_i\beta}}{\sum_{j\in R_i} e^{x_j\beta}} \quad \text{(cf. (2.19))}$$

for estimating the parameter β, where $x_i \in \mathbb{R}^p$ and y_i are covariates and time, respectively, and $\delta_i = 1$ and $\delta_i = 0$ correspond to death and survey termination, respectively, for $i = 1, \ldots, N$.

On the other hand, R_i is the set of indices j such that $y_j \ge y_i$. We formulate the Lasso as follows:

$$-\frac{1}{N} \sum_{i:\delta_i=1} \log \frac{e^{x_i\beta}}{\sum_{j\in R_i} e^{x_j\beta}} + \lambda\|\beta\|_1 . \quad \text{(cf. (2.20))}$$

To solve the solution, similar to logistic and Poisson regressions, we compute u, W, where L is defined as follows:

$$L := - \sum_{i:\delta_i=1} \log \frac{e^{x_i\beta}}{\sum_{j\in R_i} e^{x_j\beta}} .$$

(a) Let $j \in R_i$, $\delta_i = 1 \iff i \in C_j$. Show

$$\frac{\partial L}{\partial \beta_k} = -\sum_{i:\delta_i=1} \left\{ x_{i,k} - \frac{\sum_{j\in R_i} x_{j,k} e^{x_j \beta}}{\sum_{h\in R_i} e^{x_h \beta}} \right\}$$

$$= -\sum_{i=1}^{N} x_{i,k} \left\{ \delta_i - \sum_{j\in C_i} \frac{e^{x_i \beta}}{\sum_{h\in R_j} e^{x_h \beta}} \right\} .$$

Express u in $\nabla L = -X^T u$.

Hint   For $S_i = \sum_{h\in R_i} e^{x_h \beta}$, we have

$$\sum_{i:\delta_i=1} \sum_{j\in R_i} \frac{x_{j,k} e^{x_j \beta}}{S_i} = \sum_{j=1}^{N} \sum_{i\in C_j} \frac{x_{j,k} e^{x_j \beta}}{S_i} = \sum_{i=1}^{N} x_{i,k} \sum_{j\in C_i} \frac{e^{x_i \beta}}{S_j} .$$

(b) Each element of $\nabla^2 L = X^T W X$ can be expressed by

$$\frac{\partial^2 L}{\partial \beta_k \partial \beta_l} = \sum_{i=1}^{N} x_{i,k} \sum_{j\in C_i} \frac{\partial}{\partial \beta_l} \left(\frac{e^{x_i \beta}}{\sum_{h\in R_j} e^{x_h \beta}} \right)$$

$$= \sum_{i=1}^{N} x_{i,k} \sum_{j\in C_i} \frac{1}{(\sum_{r\in R_j} e^{x_r \beta})^2} \{ x_{i,l} e^{x_i \beta} \sum_{s\in R_j} e^{x_s \beta} - e^{x_i \beta} \sum_{h\in R_j} x_{h,l} e^{x_h \beta} \}$$

$$= \sum_{i=1}^{N} \sum_{h=1}^{N} x_{i,k} x_{h,l} \sum_{j\in C_i} \frac{e^{x_i \beta}}{(\sum_{r\in R_j} e^{x_r \beta})^2} \{ I(i=h) \sum_{s\in R_j} e^{x_s \beta} - I(h\in R_j) e^{x_h \beta} \} .$$

Find the diagonal elements of W.

Hint   When $i = h$, $j \in C_i$ implies $h \in R_j$.

32. For the data 1846-2568-2-SP.rda (Alizadeh, 2000) $https : //www.jstatsoft.org/rt/suppFileMetadata/v039i05/0/722$ of malignant lymphoma survival time, answer the following:

(a) Download the data and execute the following procedure:

```
library(survival)
load("LymphomaData.rda"); attach("LymphomaData.rda")
names(patient.data); x = t(patient.data$x)
y = patient.data$time; delta = patient.data$status; Surv(y,
    delta)
```

The variables x, y, and delta store the expression data of $p = 7{,}399$ genes, times, and death ($= 1$) or survey termination ($= 0$). The sample size is $N = 240$. Execute the following code, find the λ that minimizes the CV, and find how many genes have nonzero coefficients among the 7,399. Moreover, output cv.fit.

```
cv.fit = cv.glmnet(x, Surv(y, delta), family = "cox")
```

(b) Fill in the blank and draw the Kaplan-Meier curve for samples such that $x_i \beta > 0$ and for those such that $x_i \beta < 0$ to distinguish them.

```
1  fit2 = glmnet(x, Surv(y, delta), lambda = cv.fit$lambda.min,
       family = "cox")
2  z = sign(drop(x %*% fit2$beta))
3  fit3 = survfit(Surv(y, delta) ~ ## Blank ##)
4  autoplot(fit3)
5  mean(y[z == 1])
6  mean(y[z == -1])
```

33. It is assumed that logistic regression and the support vector machine (SVM) are similar even when Lasso is applied.

    a. The following code executes Lasso for logistic regression and the SVM for the South Africa heart disease dataset: https://www2.stat.duke.edu/~cr173/Sta102_Sp14/Project/heart.pdf.

    Because those procedures are in separate packages glmnet and sparseSVM, the plots are different, and no legend is available for the first plot. Thus, we construct a graph using the coefficients output by the packages. Output the graph for the SVM as well.

```
1   library(ElemStatLearn)
2   library(glmnet)
3   library(sparseSVM)
4   data(SAheart)
5   df = SAheart
6   df[, 5] = as.numeric(df[, 5])
7   x = as.matrix(df[, 1:9]); y = as.vector(df[, 10])
8   p = 9
9   binom.fit = glmnet(x, y, family = "binomial")
10  svm.fit = sparseSVM(x, y)
11  par(mfrow = c(1, 2))
12  plot(binom.fit); plot(svm.fit, xvar = "norm")
13  par(mfrow = c(1, 1))
14  ## The outputs seemed  to be similar, but we are not convinced
          that they are close because no legend is available.
15  ## So, we made a graph from scratch.
16  coef.binom = binom.fit$beta; coef.svm = coef(svm.fit)[2:(p +
          1), ]
17  norm.binom = apply(abs(coef.binom), 2, sum); norm.binom = norm
          .binom / max(norm.binom)
18  norm.svm = apply(abs(coef.svm), 2, sum); norm.svm = norm.svm /
          max(norm.svm)
19  par(mfrow = c(1, 2))
20  plot(norm.binom, xlim = c(0, 1), ylim = c(min(coef.binom), max
          (coef.binom)),
21      main = "Logistic Regression", xlab = "Norm", ylab = "
          Coefficient", type = "n")
22  for (i in 1:p) lines(norm.binom, coef.binom[i, ], col = i)
23  legend("topleft", legend = colnames(df), col = 1:p, lwd = 2,
          cex = .8)
24  par(mfrow = c(1, 1))
```

    b. From the gene expression data of past patients with leukemia, we distinguish between acute lymphocytic leukemia (ALL) and acute myeloid leukemia

(AML) for each future patient. In particular, our goal is to determine which genes should be checked to distinguish between them. To this end, we download the training data file leukemia_big.csv from the site listed below: $https://web.stanford.edu/\ hastie/CASI_files/DATA/leukemia.html$. The data contain samples for $N = 72$ patients (47 ALL and 25 AML) and $p = 7{,}128$ genes: $https://www.ncc.go.jp/jp/rcc/about/pediatric_leukemia/index.html$.

After executing the following, output the coefficients obtained via logistic regression and the SVM. In most of the genome data, the rows and columns are genes and samples, respectively. However, similar to the dataset we have seen thus far, the rows and columns are instead samples and genes.

```
1  df = read.csv("http://web.stanford.edu/~hastie/CASI_files/DATA
        /leukemia_big.csv")
2  dim(df)
3  names = colnames(df)
4  x = t(as.matrix(df))
5  y = as.numeric(substr(names, 1, 3) == "ALL")
6  p = 7128
7  binom.fit = glmnet(x, y, family = "binomial")
8  svm.fit = sparseSVM(x, y)
9  coef.binom = binom.fit$beta; coef.svm = coef(svm.fit)[2:(p +
        1), ]
10 norm.binom = apply(abs(coef.binom), 2, sum); norm.binom = norm
        .binom / max(norm.binom)
11 norm.svm = apply(abs(coef.svm), 2, sum); norm.svm = norm.svm /
        max(norm.svm)
```

Chapter 3
Group Lasso

Group Lasso is Lasso such that the variables are categorized into K groups $k = 1, \ldots, K$. The p_k variables $\theta_k = [\theta_{1,k}, \ldots, \theta_{p_k,k}]^T \in \mathbb{R}^{p_k}$ in the same group share the same times at which the nonzero coefficients become zeros when we increase the λ value. This chapter considers groups with nonzero and zero coefficients to be active and nonactive, respectively, for each λ. In other words, group Lasso chooses active groups rather than active variables. The active and nonactive status may be different among the groups.

Example 24 (**Multiple Responses**) For the linear regression discussed in Chap. 1, we consider only one response. However, in this section, we also consider the case where K responses exist. If there are p covariates, we require pK coefficients. We consider the situation such that for each $j = 1, \ldots, p$, the K coefficients share the active/nonactive status [27]. For example, suppose that in a baseball league, the numbers of HRs (home runs) and RBIs (runs batted in) are responses and the numbers of hits and walks are covariates. Then, the correlation between HRs and RBIs is strong, and we anticipate that the solution paths of $p = 2$ covariates over $\lambda \geq 0$ are similar between the $K = 2$ coefficients. We discuss the problem in Sect. 3.6.

Example 25 (**Logistic Regression for Multiple Values**) For the logistic regression and classification discussed in Chap. 2, we might have constructed a group consisting of $k = 1, \ldots, K$ for each covariate $j = 1, \ldots, p$ to determine which covariate plays an important role in the classification task [27]. For example, in Fisher's Iris dataset, although there are $p \times K = 4 \times 3 = 12$ parameters, we may divide them into $p = 4$ (sepal/petal, width/length) groups that contain $K = 3$ iris species to observe the $p = 4$ active/nonactive status. When we increase λ, $K = 3$ coefficients in the same variable indexed by j become zero at once for some $\lambda > 0$. We discuss the problem in Sect. 3.7.

© The Author(s), under exclusive license to Springer Nature Singapore Pte Ltd. 2021 77
J. Suzuki, *Sparse Estimation with Math and R*,
https://doi.org/10.1007/978-981-16-1446-0_3

Example 26 (**Generalized Additive Model**) Similar to the linear regression discussed in Chap. 1, we are given the data $X \in \mathbb{R}^{N \times p}$, $y \in \mathbb{R}^N$.

We consider the case where for each $i = 1, \ldots, N$, we can write

$$y_i = \sum_{k=1}^{K} f_k(x_i; \theta_k) + \epsilon_i \,,$$

where f_k contains p_k parameters $\theta_k = [\theta_{1,k}, \ldots, \theta_{p_k,k}] \in \mathbb{R}^{p_k}$ and the noise ϵ_i follows a Gaussian distribution with zero mean and unknown variance $\sigma^2 > 0$. If we apply the problem to Lasso, because we decide whether each function f_k should be contained, we need to decide whether each θ_k rather than each of $\theta_{j,k}$, $j = 1, \ldots, p_k$, should be active as a group [23]. We discuss the problem in Sect. 3.8.

In this chapter, generalizing the notion of Lasso, we categorize the variables into K groups, each of which contains p_k variables ($k = 1, \ldots, K$), and given observations $z_{i,k} \in \mathbb{R}^{p_k}$ ($k = 1, \ldots, K$), $y_i \in \mathbb{R}$ ($i = 1, \ldots, N$), we consider the problem of finding the

$$\theta_1 = [\theta_{1,1}, \ldots, \theta_{p_1,1}]^T, \ldots, \theta_K = [\theta_{1,K}, \ldots, \theta_{p_K,K}]^T$$

that minimize

$$\frac{1}{2} \sum_{i=1}^{N} (y_i - \sum_{k=1}^{K} z_{i,k} \theta_k)^2 + \lambda \sum_{k=1}^{K} \|\theta_k\|_2 \tag{3.1}$$

Although similar to the previous chapters, we require data preprocessing such as centralization and normalization in this chapter. In addition, we skip the details when we deal with numerical data generated according to the standard Gaussian distribution, except in Sect. 3.6. We write $\|\theta_k\|_2 := \sqrt{\sum_{j=1}^{p_k} \theta_{j,k}^2}$ and regard $z_{i,k}$ as a row vector.

Note that (3.1) is a generalization of the linear Lasso ($p_1 = \cdots = p_K = 1, K = p$) discussed in Chap. 1. In fact, if we let $x_i := [z_{i,1}, \ldots, z_{i,p}] \in \mathbb{R}^{1 \times p}$ (row vector) and $\beta := [\theta_{1,1}, \ldots, \theta_{p,1}]^T$, we have

$$\frac{1}{2} \sum_{i=1}^{N} (y_i - x_i \beta)^2 + \lambda \|\beta\|_1 \,.$$

3.1 When One Group Exists

In this section, we consider the case where only one group exists $K = 1$ and $p_k = p$. Then, (3.1) can be expressed by

$$\frac{1}{2}\sum_{i=1}^{N}(y_i - z_{i,1}\theta_1)^2 + \lambda\|\theta_1\|_2 \ .$$

We proceed the discussion by replacing p_1 with p, $z_{i,1} \in \mathbb{R}^p$ with $x_i = [x_{i,1}, \dots, x_{i,p}]$ (row vector), and $\theta_1 \in \mathbb{R}^p$ with $\beta = [\beta_1, \dots, \beta_p]^T$. In particular, we find the optimum β via a method different from coordinate descent. Although the second term was $\lambda\|\beta\|_1 = \lambda\sum_{j=1}^{p}|\beta_j|$ in Chap. 1, it is $\lambda\|\beta\|_2 = \lambda\sqrt{\sum_{j=1}^{p}\beta_j^2}$ in this section, different from that of Ridge regression $\lambda\|\beta\|_2^2 = \lambda\sum_{j=1}^{p}\beta_j^2$ as well.

For the function

$$f(x, y) := \sqrt{x^2 + y^2} \ , \tag{3.2}$$

if we set $y = 0$, because $f(x, y) = |x|$, the slopes when approaching from the left and right are different. On the other hand, the partial differentials outside the origin are

$$\begin{cases} f_x(x, y) = x/\sqrt{x^2 + y^2} \\ f_y(x, y) = y/\sqrt{x^2 + y^2} \end{cases} , \tag{3.3}$$

and they are continuous.

We define the subderivative of one-variable function in Chap. 1. Similarly, we define the subderivative of $f : \mathbb{R} \to \mathbb{R}$ at $(x_0, y_0) \in \mathbb{R}^2$ by the set of $(u, v) \in \mathbb{R}^2$ such that

$$f(x, y) \geq f(x_0, y_0) + u(x - x_0) + v(y - y_0) \tag{3.4}$$

for all $(x, y) \in \mathbb{R}^2$. Similar to the one-variable functions, if f is differentiable at (x_0, y_0), the subderivative (u, v) is the derivative at (x_0, y_0).

As we did in Chap. 1, we consider the subderivative at the origin. First, from the definition (3.4), the subderivative of (3.2) at $(x_0, y_0) = (0, 0)$ is the set of $(u, v) \in \mathbb{R}^2$ such that

$$\sqrt{x^2 + y^2} \geq ux + vy$$

for all $(x, y) \in \mathbb{R}^2$. If we let $x = r\cos\theta$, $y = r\sin\theta$ and $u = s\cos\phi$, $v = s\sin\phi$ ($s \geq 0$, $0 \leq \phi < 2\pi$), we require $r \geq rs\cos(\theta - \phi)$ for arbitrary $r \geq 0$, $0 \leq \theta < 2\pi$. Thus, if (u, v) is outside the unit circle ($s > 1$), then it is not included in the subderivative. Conversely, if it is inside the unit circle ($s \leq 1$), the inequality holds. Thus, the disk $\{(u, v) \in \mathbb{R}^2 \mid u^2 + v^2 \leq 1\}$ is the subderivative.

In the following, we consider finding the value of β that minimizes

$$\frac{1}{2}\|y - X\beta\|_2^2 + \lambda\|\beta\|_2 \tag{3.5}$$

from $X \in \mathbb{R}^{N \times p}$ and $y \in \mathbb{R}^N$, $\lambda \geq 0$, where we denote $\|\beta\|_2 := \sqrt{\sum_{j=1}^{p}\beta_j^2}$ for $\beta = [\beta_1, \dots, \beta_p]^T \in \mathbb{R}^p$. In this chapter, we do not divide the first term by N as in (3.5). We may interpret $N\lambda$ as if it was λ.

(a) $\beta \neq 0$ $(p = 1)$ (b) $\beta = 0$ $(p = 1)$

(c) $\beta \neq \begin{bmatrix} 0 \\ 0 \end{bmatrix}$ $(p = 2)$ (d) $\beta = \begin{bmatrix} 0 \\ 0 \end{bmatrix}$ $(p = 2)$

Fig. 3.1 For $p = 1$, if $X^T y$ is a distance λ from the origin, then $\beta \neq 0$. If within λ, then $\beta = 0$. For $p = 2$, if $X^T y$ is a distance λ from the origin, then $\beta \neq [0, 0]^T$. If within λ, then $\beta = [0, 0]^T$

For (3.5), if $p = 1$, the equation that gives the subdirective $\ni 0$ is written as

$$- X^T (y - X\beta) + \lambda[-1, 1] \ni 0 . \tag{3.6}$$

If $\beta \neq 0$, as we discussed in Chap. 1, under the assumption that $X^T X$ is the unit matrix, we have $\beta = \mathcal{S}_\lambda(X^T y)$. If $\beta = 0$, substituting $\beta = 0$ into (3.6), we have

$$|X^T y| \leq \lambda . \tag{3.7}$$

(3.7) means that the line segment with the center $X^T y$ of length $\pm \lambda$ contains the origin (Fig. 3.1a, b).

Suppose $p = 2$. When $(\beta_1, \beta_2) \neq (0, 0)$, from (3.3), the subderivative of $\|\beta\|_2$ at β is $\beta / \|\beta\|_2$, and the equation that gives the subderivative that contains zero becomes

$$- X^T (y - X\beta) + \lambda \frac{\beta}{\|\beta\|_2} \ni \begin{bmatrix} 0 \\ 0 \end{bmatrix} , \tag{3.8}$$

which means

$$X^T X\beta = X^T y - \lambda \frac{\beta}{\|\beta\|_2} . \tag{3.9}$$

Thus, the original point $X^T y$ approaches the origin by the length λ due to Lasso. On the other hand, when $\beta = 0$, we have

$$\text{The solution of (3.5) is } \beta = 0 \iff -X^T y + \lambda \left\{ \begin{bmatrix} u \\ v \end{bmatrix} \middle| u^2 + v^2 \leq 1 \right\} \ni \begin{bmatrix} 0 \\ 0 \end{bmatrix}$$

$$\iff \|X^T y\|_2 \leq \lambda ,$$

which means that the disk with a center and radius of $X^T y$ and λ, respectively, contains the origin (Fig. 3.1c, d).

For $p = 1$, we have the formula $\mathcal{S}_\lambda(X^T y)$. However, for $p \geq 2$, we wait for the convergence similar to the Newton method discussed in Chap. 2—Repeating a recursive equation. We consider a method for finding the solution $\beta \in \mathbb{R}^p$.

Setting $\nu > 0$, we repeat the updates

$$\gamma := \beta + \nu X^T (y - X\beta) \tag{3.10}$$

$$\beta = \left(1 - \frac{\nu \lambda}{\|\gamma\|_2} \right)_+ \gamma \tag{3.11}$$

to await the convergence of $\beta \in \mathbb{R}^p$, where we denote $(u)_+ := \max\{0, u\}$.

We give a validation of the method and specify the value of ν.

The procedure is constructed as follows.

```
gr = function(X, y, lambda) {
    nu = 1 / max(eigen(t(X) %*% X)$values)
    p = ncol(X)
    beta = rep(1, p); beta.old = rep(0, p)
    while (max(abs(beta - beta.old)) > 0.001) {
        beta.old = beta
        gamma = beta + nu * t(X) %*% (y - X %*% beta)
        beta = max(1 - lambda * nu / norm(gamma, "2"), 0) * gamma
    }
    return(beta)
}
```

Example 27 Artificially generating data, we execute the function gr. We show the result in Fig. 3.2, in which we observe that the p variables share the λ value in which the coefficients become zero when we increase λ.

```
## Data Generation
n = 100
p = 3
X = matrix(rnorm(n * p), ncol = p); beta = rnorm(p); epsilon = rnorm(n)
y = 0.1 * X %*% beta + epsilon
## Display the Change of Coefficients
lambda = seq(1, 50, 0.5)
m = length(lambda)
beta = matrix(nrow = m, ncol = p)
for (i in 1:m) {
    est = gr(X, y, lambda[i])
    for (j in 1:p) beta[i, j] = est[j]
```

```
13  }
14  y.max = max(beta); y.min = min(beta)
15  plot(lambda[1]:lambda[m], ylim = c(y.min, y.max),
16      xlab = "lambda", ylab = "Coefficients", type = "n")
17  for (j in 1:p) lines(lambda, beta[, j], col = j + 1)
18  legend("topright", legend = paste("Coefficients", 1:p), lwd = 2, col =
          2:(p + 1))
19  segments(lambda[1], 0, lambda[m], 0)
```

3.2 Proximal Gradient Method

When we alternately update (3.10) and (3.11), the sequence $\{\beta_t\}$ is not guaranteed to converge to the solution of (3.5). In the following, we show that we can always obtain the correct solution by choosing an appropriate value of $\nu > 0$.

Suppose $\lambda = 0$. Then, we note that the update via (3.10) and (3.11) amounts to

$$\beta_{t+1} \leftarrow \beta_t + \nu X^T(y - X\beta_t) \quad (t = 0, 1, \ldots)$$

after setting an initial value β_0. Let $g(\beta)$ be (3.5) with $\lambda = 0$. Then, since $-X^T(y - X\beta)$ is the derivative of $g(\beta)$ by β, we can write the update as

$$\beta_{t+1} \leftarrow \beta_t - \nu \nabla g(\beta_t) . \tag{3.12}$$

We call the procedure a gradient method because it updates β such that the decrease in g is maximized when it minimizes g.

In the following, we express the second term of (3.5) by $h(\beta) := \lambda\|\beta\|_2$ and evaluate the convergence of a variant of the gradient method (proximal gradient)

$$\beta_{t+1} \leftarrow \beta_t - \nu\{\nabla g(\beta_t) + \partial h(\beta)\}$$

for $\lambda > 0$, where ∂ denotes the subgradient. If we use the function

$$\text{prox}_h(z) := \arg\min_{\theta \in \mathbb{R}^p} \left\{ \frac{1}{2}\|z - \theta\|_2^2 + h(\theta) \right\}$$

(proxy operation), we can express the update as

$$\beta_{t+1} \leftarrow \text{prox}_{\nu h}(\beta_t - \nu \nabla g(\beta_t)) . \tag{3.13}$$

In fact, (3.13) substitutes into β_{t+1} the $\theta \in \mathbb{R}^p$ that minimizes

$$\frac{1}{2}\|\beta_t - \nu\nabla g(\beta_t) - \theta\|_2^2 + \nu h(\theta) , \tag{3.14}$$

which can be examined by differentiating (3.14) by θ.

The above procedure gives a general method that finds the minimum value of the function f expressed by the sum of convex functions g, h with g differentiable. We can apply (3.10) and (3.11) to the standard optimization problem, in which γ and β are updated via $\gamma \leftarrow \beta_t - \nu \nabla g(\beta_t)$ and $\beta \leftarrow \text{prox}_{\nu h}(\gamma)$, respectively.

Finally, by repeatedly applying (3.13), we find an appropriate ν by which we obtain $\beta = \beta^*$ that minimizes $f = g + h : \mathbb{R}^p \to \mathbb{R}$. We note that there exists a constant L such that

$$(x - y)^T \nabla^2 g(z)(x - y) \le L\|x - y\|_2^2 \tag{3.15}$$

for arbitrary $x, y, z \in \mathbb{R}^p$. We refer to this L as a Lipschitz constant . In fact, if an orthogonal matrix P diagonalizes $X^T X$, from $\nabla^2 g(z) = X^T X$, $P^T (LI - X^T X) P$ is nonnegative definite, and the maximum eigenvalue of $X^T X$ is a Lipschitz constant.

In the following, we make the following general assumption: g, h are convex, g is differentiable, and $L > 0$ is a Lipschitz constant. We define

$$Q(x, y) := g(y) + (x - y)^T \nabla g(y) + \frac{L}{2}\|x - y\|^2 + h(x) \tag{3.16}$$

$$p(y) := \text{argmin}_{x \in \mathbb{R}^p} Q(x, y) \tag{3.17}$$

for arbitrary $x, y \in \mathbb{R}^p$. Then, the update (3.13) can be expressed by

$$\beta_{t+1} \leftarrow p(\beta_t) . \tag{3.18}$$

This method is called the ISTA (iterative shrinkage-thresholding algorithm), and we can show that the precision of the procedure $\beta_{t+1} \leftarrow p(\beta_t)$ is at most $O(k^{-1})$ for k repetitions.

Let $\nu := 1/L$.

Proposition 7 (Beck and Teboulle, 2009 [3]) *The sequence $\{\beta_t\}$ generated by the ISTA satisfies*

$$f(\beta_k) - f(\beta_*) \le \frac{L\|\beta_0 - \beta_*\|_2^2}{2k} ,$$

where β_* is the optimum solution.

For the proof, see the Appendix.

Next, we modify the ISTA. Using the sequence $\{\alpha_t\}$

$$\alpha_1 = 1, \quad \alpha_{t+1} := \frac{1 + \sqrt{1 + 4\alpha_t^2}}{2} \tag{3.19}$$

we generate $\{\gamma_t\}$ as well as $\{\beta_t\}$: setting $\gamma_1 = \beta_0 \in \mathbb{R}^p$, we update the sequences via $\beta_t = p(\gamma_t)$ and

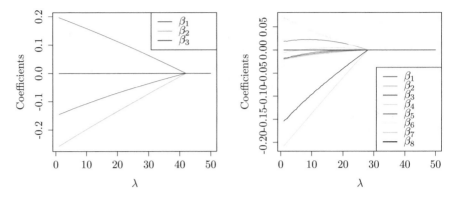

Fig. 3.2 We execute the function `gr` (Example 27) for $p = 3$ and $p = 8$ and observe that the coefficients become zero at the same time as we increase λ

$$\gamma_{t+1} = \beta_t + \frac{\alpha_t - 1}{\alpha_{t+1}}(\beta_t - \beta_{t-1}) . \qquad (3.20)$$

This method is called the FISTA (fast iterative shrinkage-thresholding algorithm). Then, we achieve an improvement in the performance over that of the ISTA. The convergence rate is at most $O(k^{-2})$ for k repetitions.

Proposition 8 (Beck and Teboulle, 2009 [3]) *The sequence $\{\beta_t\}$ generated by the FISTA satisfies*

$$f(\beta_k) - f(\beta_*) \leq \frac{L\|\beta_0 - \beta_*\|_2^2}{(k + 1)^2} ,$$

where β_* is the optimum value.[1]

For the proof, see Page 193 of Beck and Teboulle (2009). Using Lemma 1, Proposition 8 can be obtained from lengthy but plain formula transformations. If we replace the function `gr` in the group Lasso with the FISTA, we obtain the following procedure. If we execute the same procedure as in Example 27, we obtain the same shape as in Fig. 3.2.

```
1  fista = function(X, y, lambda) {
2    nu = 1 / max(eigen(t(X) %*% X)$values)
3    p = ncol(X)
4    alpha = 1
5    beta = rep(1, p); beta.old = rep(1, p)
6    gamma = beta
7    while (max(abs(beta - beta.old)) > 0.001) {
8      print(beta)
9      beta.old = beta
10     w = gamma + nu * t(X) %*% (y - X %*% gamma)
11     beta = max(1 - lambda * nu / norm(w, "2"), 0) * w
```

---

[1] Nesterov's accelerated gradient method (2007) [22].

```
12    alpha.old = alpha
13    alpha = (1 + sqrt(1 + 4 * alpha ^ 2)) / 2
14    gamma = beta + (alpha.old - 1) / alpha * (beta - beta.old)
15    }
16    return(beta)
17  }
```

Example 28 To compare the efficiencies of the ISTA and FISTA, we construct the program (Fig. 3.3). We could not see a significant difference when $N = 100$, $p = 1, 3$. However, the FISTA can be applied to general optimization problems and works efficiently for large-scale problems.[2] We show the program in Exercise 39.

3.3 Group Lasso

In the following, we apply the group Lasso procedure for one group in a cyclic manner to obtain the group Lasso for $K \geq 1$ groups. We can obtain the solution of (3.1) using the coordinate descent method among the groups. If each of the groups $k = 1, \ldots, K$ contains p_k variables, we apply the method introduced in Sects. 3.1 and 3.2 as if $p = p_k$.

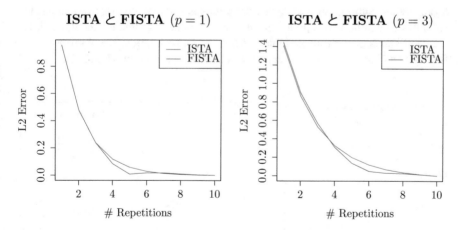

Fig. 3.3 The difference in the convergence rate between the ISTA and FISTA (Example 28). It seems that the difference is not significant unless the problem size is rather large

---

[2] If there exists $m > 0$ such that $(\nabla f(x) - \nabla f(y))^T (x - y) \geq \frac{m}{2} \|x - y\|^2$ for arbitrary $x, y \in \mathbb{R}$, then we say that $f : \mathbb{R} \to \mathbb{R}$ is strongly convex. In this case, the error can be exponentially decreased w.r.t. k (Nesterov, 2007).

When we apply the coordinate decent method, we compute the residual

$$r_{i,k} = y_i - \sum_{h \neq k} z_{i,h} \theta_h \quad (i = 1, \ldots, N) .$$

For example, we can construct the following procedure.

```
1   group.lasso = function(z, y, lambda = 0) {
2     J = length(z)
3     theta = list(); for (j in 1:J) theta[[j]] = rep(0, ncol(z[[j]]))
4     for (m in 1:10) {
5       for (j in 1:J) {
6         r = y; for (k in 1:J) {if (k != j) r = r - z[[k]] %*% theta[[k]]}
7         theta[[j]] = gr(z[[j]], r, lambda)   # fista(X, r, lambda) is ok
8       }
9     }
10    return(theta)
11  }
```

Note that z and theta are lists. If the number of groups is one, they are the X, β discussed in Chap. 1.

For a single group, we have seen that the variables share the active/nonactive status for $\lambda \geq 0$. Variables in the same group share the same status for multiple groups, but those in different groups do not. We manually input the variables into the procedure for grouping before executing it (the grouping is not automatically constructed).

Example 29 Suppose that we have four variables and that the first two and last two variables constitute two separate groups. We artificially generate data and estimate the coefficients of the four variables. When we make λ larger, we observe that the first two and last two variables share the active/nonactive status (Fig. 3.4).

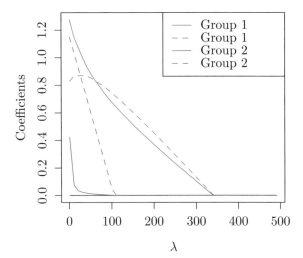

Fig. 3.4 Group Lasso with $J = 2$, $p_1 = p_2 = 2$ (Example 29). If we make λ larger, we observe that each group shares the active/nonactive status

```
1    ## Data Generation
2    N = 100; J = 2
3    u = rnorm(n); v = u + rnorm(n)
4    s = 0.1 * rnorm(n); t = 0.1 * s + rnorm(n); y = u + v + s + t + rnorm(n)
5    z = list(); z[[1]] = cbind(u, v); z[[2]] = cbind(s, t)
6    ## Display the coefficients that change with lambda
7    lambda = seq(1, 500, 10); m = length(lambda); beta = matrix(nrow = m,
          ncol = 4)
8    for (i in 1:m) {
9      est = group.lasso(z, y, lambda[i])
10     beta[i, ] = c(est[[1]][1], est[[1]][2], est[[2]][1], est[[2]][2])
11   }
12   y.max = max(beta); y.min = min(beta)
13   plot(lambda[1]:lambda[m], ylim = c(y.min, y.max),
14         xlab = "lambda", ylab = "Coefficients", type = "n")
15   lines(lambda, beta[, 1], lty = 1, col = 2); lines(lambda, beta[, 2], lty
          = 2, col = 2)
16   lines(lambda, beta[, 3], lty = 1, col = 4); lines(lambda, beta[, 4], lty
          = 2, col = 4)
17   legend("topright", legend = c("Group1", "Group1", "Group2", "Group2"),
18          lwd = 1, lty = c(1, 2), col = c(2, 2, 4, 4))
19   segments(lambda[1], 0, lambda[m], 0)
```

3.4 Sparse Group Lasso

In the following, we extend the formulation of (3.1) to

$$\frac{1}{2} \sum_{i=1}^{N} (y_i - \sum_{k=1}^{K} z_{i,k}\theta_k)^2 + \lambda \sum_{k=1}^{K} \{(1 - \alpha)\|\theta_k\|_2 + \alpha\|\theta_k\|_1\} \tag{3.21}$$

to contain sparsity within the group as well as between groups [26]. We refer to this extended group Lasso as sparse group Lasso. The difference lies only in the second term from (3.1), introducing the parameter $0 < \alpha < 1$, as done in the elastic net.

In sparse group Lasso, although the variables of the nonactive groups are non-active, the variables in the active groups may be either active or nonactive. In this sense, sparse group Lasso extends the ordinary group Lasso and allows the active groups to have nonactive variables.

If we differentiate both sides of (3.21) by $\theta_k \in \mathbb{R}^{p_k}$, we have

$$-\sum_{i=1}^{N} z_{i,k}(r_{i,k} - z_{i,k}\theta_k) + \lambda(1 - \alpha)s_k + \lambda\alpha t_k = 0 \tag{3.22}$$

with $r_{i,k} := y_i - \sum_{l \neq k} z_{i,l}\hat{\theta}_l$, where $s_k, t_k \in \mathbb{R}^{p_k}$ are the subderivatives of $\|\theta_k\|_2$, $\|\theta_k\|_1$.

Next, we derive the conditions for $\theta_k = 0$ to be optimum in terms of s_k, t_k.

The minimum value of (3.21) except the term with the coefficient $1 - \alpha$ is

$$\theta_k = \mathcal{S}_{\lambda\alpha} \left(\sum_{i=1}^{N} z_{i,k} r_{i,k} \right) \Big/ \sum_{i=1}^{N} z_{i,k}^2 \,,$$

which means that a necessary for $\theta_k = 0$ to have a solution in (3.22) is

$$\lambda(1 - \alpha) \geq \left\| \mathcal{S}_{\lambda\alpha} \left(\sum_{i=1}^{N} z_{i,k} r_{i,k} \right) \right\|_2 .$$

Then, (3.11), which holds for $\alpha = 0$, is extended to

$$\beta = \left(1 - \frac{\nu\lambda(1 - \alpha)}{\|\mathcal{S}_{\lambda\alpha}(\gamma)\|_2} \right)_{+} \mathcal{S}_{\lambda\alpha}(\gamma) \,, \tag{3.23}$$

where $(u)_+ := \max\{0, u\}$. Therefore, we can construct a procedure that repeats (3.10) and (3.23) alternatively.

Although we abbreviate the proof, we have the following proposition.

Proposition 9 *For arbitrary $0 < \nu < \frac{1}{L}$, $\beta \in \mathbb{R}^p$ minimizing (3.21) and $\beta \in \mathbb{R}^p$ being a solution of (3.10), (3.23) are equivalent.*

The actual procedure can be constructed as follows, where the three lines marked as ## are different from the ordinary group Lasso.

```
sparse.group.lasso = function(z, y, lambda = 0, alpha = 0) {
  J = length(z)
  theta = list(); for (j in 1:J) theta[[j]] = rep(0, ncol(z[[j]]))
  for (m in 1:10) {
    for (j in 1:J) {
      r = y; for (k in 1:J) {if (k != j) r = r - z[[k]] %*% theta[[k]]}
      theta[[j]] = sparse.gr(z[[j]], r, lambda, alpha)
          ##
    }
  }
  return(theta)
}

sparse.gr = function(X, y, lambda, alpha = 0) {
  nu = 1 / max(2 * eigen(t(X) %*% X)$values)
  p = ncol(X)
  beta = rnorm(p); beta.old = rnorm(p)
  while (max(abs(beta - beta.old)) > 0.001) {
    beta.old = beta
    gamma = beta + nu * t(X) %*% (y - X %*% beta)
    delta = soft.th(lambda * alpha, gamma)
          ##
    beta = max(1 - lambda * nu * (1 - alpha) / norm(delta, "2"), 0) *
      delta    ##
  }
  return(beta)
}
```

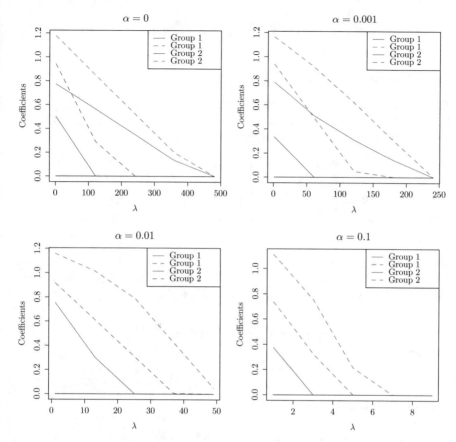

Fig. 3.5 We observe that the larger the value of α, the greater the number of variables that are active/nonactive in the group (Example 30)

Example 30 We executed the sparse group Lasso for $\alpha = 0, 0.001, 0.01, 0.1$ under the same setting as in Example 29. We find that the larger the value of α, the greater the number of variables that are nonactive in the active groups (Fig. 3.5).

3.5 Overlap Lasso

We consider the case where some groups overlap. For example, the groups are $\{1, 2, 3\}, \{3, 4, 5\}$ rather than $\{1, 2\}, \{3, 4\}, \{5\}$.

For groups $k = 1, \ldots, K$, among the coefficients $\beta_1, \ldots, \beta_p$ of variables, we specify the β_k for which we do not use zero. We prepare the variables $\theta_1, \ldots, \theta_K \in \mathbb{R}^p$ such that $\mathbb{R}^p \ni \beta = \sum_{k=1}^{K} \theta_k$ (see Example 31) and find the β value that minimizes

$$\frac{1}{2}\|y - X\sum_{k=1}^{K}\theta_k\|_2^2 + \lambda\sum_{k=1}^{K}\|\theta_k\|_2$$

from the data $X \in \mathbb{R}^{N \times p}$, $y \in \mathbb{R}^N$.

Example 31 Suppose that we have five variables $\beta_1, \ldots, \beta_5$ and two groups θ_1, θ_2 that only the third variable with the coefficient β_3 may be overlapping, and that θ_1 and θ_2 contain β_1, β_2 and β_4, β_5, respectively. We write

$$\theta_1 = \begin{bmatrix} \beta_1 \\ \beta_2 \\ \beta_{3,1} \\ 0 \\ 0 \end{bmatrix}, \quad \theta_2 = \begin{bmatrix} 0 \\ 0 \\ \beta_{3,2} \\ \beta_4 \\ \beta_5 \end{bmatrix} \quad (\beta_3 = \beta_{3,1} + \beta_{3,2}),$$

and differentiate L by $\beta_1, \beta_2, \beta_{3,1}, \beta_{3,2}, \beta_4, \beta_5$. We write the first and last three columns of $X \in \mathbb{R}^{N \times 5}$ as $X_1 \in \mathbb{R}^{N \times 3}$ and $X_2 \in \mathbb{R}^{N \times 3}$. If we differentiate L by γ_1 and γ_2 (the first and last three elements of θ_1, θ_2), we have

$$\frac{\partial L}{\partial \gamma_1} = -X_1^T(y - X_1\gamma_1) + \lambda\frac{\gamma_1}{\|\gamma_1\|_2}$$

$$\frac{\partial L}{\partial \gamma_2} = -X_2^T(y - X_2\gamma_2) + \lambda\frac{\gamma_2}{\|\gamma_2\|_2}.$$

Thus, (3.8) amounts to

$$-X_1^T(y - X\theta_1) + \lambda\frac{\text{(the first 3 of } \theta_1)}{\|\theta_1\|_2} = 0$$

$$-X_2^T(y - X\theta_2) + \lambda\frac{\text{(the last 3 of } \theta_2)}{\|\theta_2\|_2} = 0.$$

If we set $\theta_j = 0$, we have $\|X_1^T y\|_2 \leq \lambda$ and $\|X_2^T y\|_2 \leq \lambda$, which means that we can optimize θ_1 and θ_2 independently.

3.6 Group Lasso with Multiple Responses

From the observations $X \in \mathbb{R}^{N \times p}$, $\beta \in \mathbb{R}^{p \times K}$, $y \in \mathbb{R}^{N \times K}$, we find the $\beta \in \mathbb{R}^{p \times K}$ such that

$$L(\beta) := L_0(\beta) + \lambda\sum_{j=1}^{p}\|\beta_j\|_2$$

with

$$L_0(\beta) := \frac{1}{2} \sum_{i=1}^{N} \sum_{k=1}^{K} (y_{i,k} - \sum_{j=1}^{p} x_{i,j}\beta_{j,k})^2 . \tag{3.24}$$

In Chap. 1, we consider only the case of $K = 1$. If we extend the number of responses to K as $y_i = [y_{i,1}, \ldots, y_{i,K}]$, we assume that for $j = 1, \ldots, p$, the K coefficients $\beta_j = [\beta_{j,1}, \ldots, \beta_{j,K}]$ (row vector) share the active/nonactive status. Setting $r_{i,k}^{(j)} := y_{i,k} - \sum_{h \neq j} x_{i,h}\beta_{h,k}$, if we differentiate (3.24) by $\beta_{j,k}$, we have

$$\sum_{i=1}^{N} \{-x_{i,j}(r_{i,k}^{(j)} - x_{i,j}\beta_{j,k})\} ,$$

which means that if we differentiate $L(\beta)$ by β_j, we have

$$\beta_j \sum_{i=1}^{N} x_{i,j}^2 - \sum_{i=1}^{N} x_{i,j}r_i^{(j)} + \lambda\partial\|\beta_j\|_2 ,$$

where $r_i^{(j)} := [r_{i,1}^{(j)}, \ldots, r_{i,K}^{(j)}]$. Therefore, the solution is

$$\hat{\beta}_j = \frac{1}{\sum_{i=1}^{N} x_{i,j}^2} \left(1 - \frac{\lambda}{\|\sum_{i=1}^{N} x_{i,j}r_i^{(j)}\|_2}\right)_+ \sum_{i=1}^{N} x_{i,j}r_i^{(j)} . \tag{3.25}$$

From the above discussion, if we consider centralization and normalization, we can construct the following procedure.

```r
gr.multi.linear.lasso = function(X, Y, lambda) {
  n = nrow(X); p = ncol(X); K = ncol(Y)
  ## Centralization　 Å¼Åšthe function centralize is defined in Chapter 1
  res = centralize(X, Y)
  X = res$X
  Y = res$y
  ## Computation of Coefficients
  beta = matrix(rnorm(p * K), p, K); gamma = matrix(0, p, K)
  while (norm(beta - gamma, "F") / norm(beta, "F") > 10 ^ (-2)) {
    gamma = beta          ## Store the beta values (for comparizon)
    R = Y - X %*% beta
    for (j in 1:p) {
      r = R + as.matrix(X[, j]) %*% t(beta[j, ])
      M = t(X[, j]) %*% r
      beta[j, ] = sum(X[, j] ^ 2) ^ (-1) * max(1 - lambda / sqrt(sum(M ^
      2)), 0) * M
      R = r - as.matrix(X[, j]) %*% t(beta[j, ])
    }
  }
  ## Intercept
  for (j in 1:p) beta[j, ] = beta[j, ] / res$X.sd[j]
  beta.0 = res$y.bar - as.vector(res$X.bar %*% beta)
  return(rbind(beta.0, beta))
}
```

Fig. 3.6 We find the relevant variables that affect HRs and RBIs via group Lasso. For each color, the solid and dotted lines express HRs and RBIs, respectively. SHs have negative coefficients, which shows that SHs and (HBPs, RBIs) have negative correlations. Even if we make λ large enough, the coefficients of H and BB do not become zero, keeping positive values, which means that they are essentially important items for HRs and RBIs

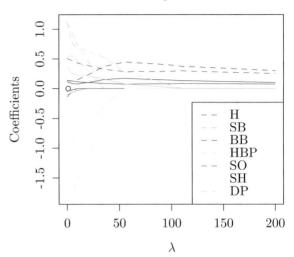

We do not have to repeat updates for each j (for each group) in this procedure but may update only once.

Example 32 From a dataset that contains nine items of a professional baseball team,[3] i.e., the number of hits (H), home runs (HRs), runs batted in (RBIs), stolen bases (SBs), base on balls (BB), hits by pitch (HBPs), strike outs (SOs), sacrifice hits (SHs), and double plays (DPs), we find the relevant covariates among the seven for the responses HRs and RBIs. In general, we require estimation of the coefficients $\beta_{j,k}$ ($j = 1, \ldots, 7$, $k = 1, 2$), because the HRs and RBIs are correlated, and we make $\beta_{j,1}, \beta_{j,2}$ share the active/nonactive status for each j. We draw the change in the coefficients with λ in Fig. 3.6.

```
1   df = read.csv("giants_2019.csv")
2   X = as.matrix(df[, -c(2, 3)])
3   Y = as.matrix(df[, c(2, 3)])
4   lambda.min = 0; lambda.max = 200
5   lambda.seq = seq(lambda.min, lambda.max, 5)
6   m = length(lambda.seq)
7   beta.1 = matrix(0, m, 7); beta.2 = matrix(0, m, 7)
8   j = 0
9   for (lambda in lambda.seq) {
10     j = j + 1
11     beta = gr.multi.linear.lasso(X, Y, lambda)
12     for (k in 1:7) {
13       beta.1[j, k] = beta[k + 1, 1]; beta.2[j, k] = beta[k + 1, 2]
14     }
15   }
16   beta.max = max(beta.1, beta.2); beta.min = min(beta.1, beta.2)
```

---

[3] The hitters of the Yomiuri Giants, Japan, each of which has at least one opportunity to bat.

```
17  plot(0, xlim = c(lambda.min, lambda.max), ylim = c(beta.min, beta.max),
18      xlab = "lambda", ylab = "Coefficients", main = "Hitters with many HR
        and RBI")
19  for (k in 1:7) {
20    lines(lambda.seq, beta.1[, k], lty = 1, col = k + 1)
21    lines(lambda.seq, beta.2[, k], lty = 2, col = k + 1)
22  }
23  legend("bottomright", c("H", "SB", "BB", "SH", "SO", "HBP", "DP"),
24          lty = 2, col = 2:8)
```

3.7 Group Lasso via Logistic Regression

In Chap. 2, we considered logistic regression with sparse estimation for binary and multiple values. In this section, among the pK parameters, the variables with the coefficients $\beta_{j,k}$ ($k = 1, \ldots, K$) with the same j comprise a group. Variables in the same group share the active/nonactive status, and the coordinate descent method is applied among the p groups.

In Chap. 2, we consider minimizing

$$L_0(\beta) + \lambda \sum_{j=1}^{p} \sum_{k=1}^{K} |\beta_{j,k}|$$

with

$$L_0(\beta) := \sum_{i=1}^{N} \left[\sum_{j=1}^{p} \sum_{k=1}^{K} y_{i,k} x_i \beta_{j,k} - \log\left(\sum_{h=1}^{K} \exp\left(\sum_{j=1}^{p} x_{i,j} \beta_{j,h} \right) \right) \right]$$

when $y_{i,k} = \begin{cases} 1, & y_i = k \\ 0, & y_i \neq k \end{cases}$. In this section, we replace the last term by the L2 norm and minimize

$$L(\beta) := L_0(\beta) + \lambda \sum_{j=1}^{p} \sqrt{ \sum_{k=1}^{K} \beta_{j,k}^2 }$$

to choose the relevant variables for classification. From the same discussion in Chap. 2, we have

$$\frac{\partial L_0(\beta)}{\partial \beta_{j,k}} = -\sum_{i=1}^{N} x_{i,j} (y_{i,k} - \pi_{i,k})$$

$$\frac{\partial^2 L_0(\beta)}{\partial \beta_{j,k} \partial \beta_{j',k'}} = \sum_{i=1}^{N} x_{i,j} x_{i,j'} w_{i,k,k'} ,$$

where $\pi_{i,k} = \dfrac{\exp(x_i \beta^{(k)})}{\sum_{h=1}^{K} \exp(x_i \beta^{(h)})}$ and (2.12) has been applied. Because $w_{i,k,k'}$ are constants, from the Taylor expansion at $\beta = \gamma$, we derive

$$L_0(\beta) = L_0(\gamma) - \sum_{i=1}^{N} \sum_{j=1}^{p} x_{ij}(\beta_j - \gamma_j)(y_i - \pi_i) + \frac{1}{2} \sum_{i=1}^{N} \sum_{j=1}^{p} \sum_{j'=1}^{p} x_{i,j} x_{i,j'}(\beta_j - \gamma_j) W_i (\beta_{j'} - \gamma_{j'})^T ,$$

where $y_i = [y_{i,1}, \ldots, y_{i,K}]^T$, $\pi_i = [\pi_{i,1}, \ldots, \pi_{i,K}]^T$ and $W_i = (w_{i,k,k'})$ are the constants at $\beta = \gamma$. We observe that the Lipschitz constant is at most t such that

$$(\beta - \gamma) W_i (\beta - \gamma)^T \leq t \|\beta - \gamma\|^2 ,$$

i.e., $t := 2 \max_{i,k} \pi_{i,k}(1 - \pi_{i,k})$. In fact, the matrix that is obtained by subtracting $W_i = (w_{i,k,k'})$ from the diagonal matrix of size K with the elements $2\pi_{i,k}(1 - \pi_{i,k})$ $(k = 1, \ldots, K)$ is $Q_i = (q_{i,k,k'})$, with

$$q_{i,k,k'} = \begin{cases} \pi_{i,k}(1 - \pi_{i,k'}), & k' = k \\ \pi_{i,k} \pi_{i,k'}, & k' \neq k \end{cases} ,$$

and

$$q_{i,k,k} = \pi_{i,k}(1 - \pi_{i,k}) = \pi_{i,k} \sum_{k' \neq k} \pi_{i,k'} = \sum_{k' \neq k} |q_{i,k,k'}|$$

for each $k = 1, \ldots, K$. Thus, from Proposition 3, the matrix Q_i is nonnegative definite.

Hence, from (3.16), (3.17), to minimize $L(\beta) = L_0(\beta) + \lambda \sum_{i=1}^{p} \|\beta_j\|_2$ with $\|\beta_j\|_2 = \sqrt{\sum_{k=1}^{K} \beta_{j,k}^2}$, we need only to minimize

$$-\sum_{i=1}^{N} \sum_{j=1}^{p} x_{i,j}(\beta_j - \gamma_j)(y_i - \pi_i) + \frac{t}{2} \sum_{i=1}^{N} \left\| \sum_{j=1}^{p} x_{i,j}(\beta_j - \gamma_j) \right\|_2^2 + \lambda \sum_{j=1}^{p} \|\beta_j\|_2 ,$$

in other words, to minimize[4]

$$\frac{1}{2} \sum_{i=1}^{N} \left\| \sum_{j=1}^{p} x_{i,j}(\beta_j - \gamma_j) - \frac{y_i - \pi_i}{t} \right\|_2^2 + \frac{\lambda}{t} \sum_{j=1}^{p} \|\beta_j\|_2$$

If we differentiate the first term by β_j, we have

---

[4] It is not essential to assume that β, γ are elements of $\mathbb{R}^p$ in (3.16), (3.17).

$$\sum_{i=1}^{N} \left\{ x_{i,j}^2 (\beta_j - \gamma_j) + x_{i,j} \sum_{h \neq j} x_{i,h} (\beta_h - \gamma_h) - x_{i,j} \frac{y_i - \pi_i}{t} \right\} = \beta_j \sum_{i=1}^{N} x_{i,j}^2 - \sum_{i=1}^{N} x_{i,j} r_i^{(j)} ,$$

where

$$r_i^{(j)} := \frac{y_i - \pi_i}{t} + \sum_{h=1}^{p} x_{i,h} \gamma_h - \sum_{h \neq j} x_{i,h} \beta_h . \tag{3.26}$$

Thus, β_j can be expressed by

$$\beta_j := \frac{1}{\sum_{i=1}^{N} x_{i,j}^2} \left(1 - \frac{\lambda/t}{\| \sum_{i=1}^{N} x_{i,j} r_i^{(j)} \|_2} \right)_+ \sum_{i=1}^{N} x_{i,j} r_i^{(j)} . \tag{3.27}$$

We suppose that one parameter is updated in each step from $\gamma_1, \ldots, \gamma_p$ to $\beta_1, \ldots, \beta_p$ and construct the procedure to repeat (3.26), (3.27) for $j = 1, \ldots, p$ as follows.

```
1   gr.multi.lasso = function(X, y, lambda) {
2     n = nrow(X); p = ncol(X); K = length(table(y))
3     beta = matrix(1, p, K)
4     gamma = matrix(0, p, K)
5     Y = matrix(0, n, K); for (i in 1:n) Y[i, y[i]] = 1
6     while (norm(beta - gamma, "F") > 10 ^ (-4)) {
7       gamma = beta
8       eta = X %*% beta
9       P = exp(eta); for (i in 1:n) P[i, ] = P[i, ] / sum(P[i, ])
10      t = 2 * max(P * (1 - P))
11      R = (Y - P) / t
12      for (j in 1:p) {
13        r = R + as.matrix(X[, j]) %*% t(beta[j, ])
14        M = t(X[, j]) %*% r
15        beta[j, ] = sum(X[, j] ^ 2) ^ (-1) * max(1 - lambda / t / sqrt(sum(
            M ^ 2)), 0) * M
16        R = r - as.matrix(X[, j]) %*% t(beta[j, ])
17      }
18    }
19    return(beta)
20  }
```

Example 33 For Fisher's Iris dataset, we draw the graph in which β changes with λ (Fig. 3.7). The same color expresses the same variable, and each contains $K = 3$ coefficients. We find that the sepal length plays an important role. The code is as follows.

Fig. 3.7 For Fisher's Iris data, we observed the changes in the coefficients with λ (Example 33). The same color expresses the same variable, and each consists of $K = 3$ coefficients. We found that the sepal length plays an important role

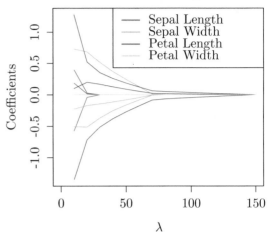

```
1   df = iris
2   X = cbind(df[[1]], df[[2]], df[[3]], df[[4]])
3   y = c(rep(1, 50), rep(2, 50), rep(3, 50))
4   lambda.seq = c(10, 20, 30, 40, 50, 60, 70, 80, 90, 100, 125, 150)
5   m = length(lambda.seq); p = ncol(X); K = length(table(y))
6   alpha = array(dim = c(m, p, K))
7   for (i in 1:m) {
8     res = gr.multi.lasso(X, y, lambda.seq[i])
9     for (j in 1:p) {for (k in 1:K) alpha[i, j, k] = res[j, k]}
10  }
11  plot(0, xlim = c(0, 150), ylim = c(min(alpha), max(alpha)), type = "n",
12       xlab = "lambda", ylab = "Coefficients", main = "Each lambda and its
        Coefficients")
13  for (j in 1:p) {for (k in 1:K) lines(lambda.seq, alpha[, j, k], col = j +
        1)}
14  legend("topright", legend = c("Sepal Length", "Sepal Width", "Petal
        Length", "Petal Width"),
15         lwd = 2, col = 2:5)
```

3.8 Group Lasso for the Generalized Additive Models

We consider Example 26 again. We prepare the basis functions $\Phi_{j,k} : \mathbb{R}^p \to \mathbb{R}$ ($j = 1, \ldots, p_k,\ k = 1, \ldots, K$) in advance. From the observations $X \in \mathbb{R}^{N \times p},\ y \in \mathbb{R}^N$, we find $\theta_{j,k}$ ($j = 1, \ldots, p_k$) that minimizes the squared error between the residual $y - \sum_{h \neq k} f_h(X) \in \mathbb{R}^N$ and

$$f_k(X; \theta_k) = \sum_{j=1}^{p_k} \theta_{j,k} \Phi_{j,k}(X)$$

for $k = 1, \ldots, K$. We repeat it over $k = 1, \ldots, K$ cyclically until convergence.

Estimating a function f in this way is called backfitting(Hastie-Tibshirani, 1990). To introduce Lasso, we formulate as follows (Ravikumar et al., 2009). For $\lambda > 0$, if we write

$$L := \frac{1}{2} \| y - \sum_{k=1}^{K} f_k(X) \|^2 + \lambda \sum_{k=1}^{K} \|\theta_k\|_2$$

$$= \frac{1}{2} \sum_{i=1}^{N} \{ y_i - \sum_{j=1}^{p_k} \sum_{k=1}^{K} \Phi_{j,k}(x_i)\theta_{j,k} \}^2 + \lambda \sum_{k=1}^{K} \|\theta_k\|_2 \qquad (3.28)$$

and $z_{i,k} := [\Phi_{1,k}(x_i), \ldots, \Phi_{p_k,k}(x_i)]$, then (3.28) coincides with (3.1).

Example 34 From the observation $(x_1, y_1), \ldots, (x_N, y_N) \in \mathbb{R} \times \mathbb{R}$, we regress y_i on $f_1(x_i) + f_2(x_i)$. Setting

$$f_1(x; \alpha, \beta) = \alpha + \beta x$$
$$f_2(x; p, q, r) = p \cos x + q \cos 2x + r \cos 3x$$

and $J = 2$, $p_1 = 2$, and $p_2 = 3$ from

$$z_1 = \begin{bmatrix} 1 & x_1 \\ \vdots & \vdots \\ 1 & x_N \end{bmatrix}, \quad z_2 = \begin{bmatrix} \cos x_1 & \cos 2x_1 & \cos 3x_1 \\ \vdots & \vdots & \vdots \\ \cos x_N & \cos 2x_N & \cos 2x_N \end{bmatrix},$$

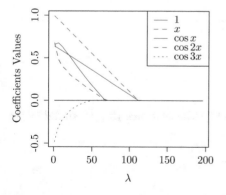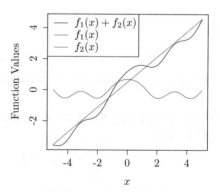

Fig. 3.8 Express $y = f(x)$ in terms of the sum of the outputs of $f_1(x)$, $f_2(x)$. We obtain the coefficients of $1, x, \cos x, \cos 2x, \cos 3x$ for each λ (left) and draw the functions $f_1(x)$, $f_2(x)$ whose coefficients have already been obtained (right)

we seek to obtain $\theta_1 = [\alpha, \beta]^T$, $\theta_2 = [p, q, r]^T$.

To this end, we made the following program and obtained the coefficients as in Fig. 3.8, left. We then decomposed the function $f(x)$ into $f_1(x)$ and $f_2(x)$ (Fig. 3.8, right).

```
1   ## Data Generation
2   n = 100; J = 2; x = rnorm(n); y = x + cos(x)
3   z[[1]] = cbind(rep(1, n), x)
4   z[[2]] = cbind(cos(x), cos(2 * x), cos(3 * x))
5   ## Display the Change of the Coefficients
6   lambda = seq(1, 200, 5); m = length(lambda); beta = matrix(nrow = m, ncol
          = 5)
7   for (i in 1:m) {
8     est = group.lasso(z, y, lambda[i])
9     beta[i, ] = c(est[[1]][1], est[[1]][2], est[[2]][1], est[[2]][2], est
          [[2]][3])
10  }
11  y.max = max(beta); y.min = min(beta)
12  plot(lambda[1]:lambda[m], ylim = c(y.min, y.max),
13        xlab = "lambda", ylab = "Coefficients", type = "n")
14  lines(lambda, beta[, 1], lty = 1, col = 2); lines(lambda, beta[, 2], lty
          = 2, col = 2)
15  lines(lambda, beta[, 3], lty = 1, col = 4); lines(lambda, beta[, 4], lty
          = 2, col = 4)
16  lines(lambda, beta[, 5], lty = 3, col = 4)
17  legend("topright", legend = c("1", "x", "cos x", "cos 2x", "cos 3x"),
18          lwd = 1, lty = c(1, 2, 1, 2, 3), col = c(2, 2, 4, 4, 4))
19  segments(lambda[1], 0, lambda[m], 0)
20
21  i = 5Ã£Â£Â£ # lambda[5] is used
22  f.1 = function(x) beta[i, 1] + beta[i, 2] * x
23  f.2 = function(x) beta[i, 3] * cos(x) + beta[i, 4] * cos(2 * x) + beta[i,
          5] * cos(3 * x)
24  f = function(x) f.1(x) + f.2(x)
25  curve(f.1(x), -5, 5, col = "red", ylab = "Function Value")
26  curve(f.2(x), -5, 5, col = "blue", add = TRUE)
27  curve(f(x), -5, 5, add = TRUE)
28  legend("topleft", legend = c("f = f.1 + f.2", "f.1", "f.2"),
29          col = c(1, "red", "blue"), lwd = 1)
```

Appendix: Proof of Proposition

Proposition 7   (Beck and Teboulle, 2009 [3]) *The sequence $\{\beta_t\}$ generated by the ISTA satisfies*

$$f(\beta_k) - f(\beta_*) \leq \frac{L\|\beta_0 - \beta_*\|_2^2}{2k} ,$$

where β_* is the optimum solution.

Proof: Before showing Proposition 7, we prove the following lemma:

Lemma 1

$$f(x) - f(p(y)) \geq \frac{L}{2} \|p(y) - y\|^2 + L(y - x)^T (p(y) - y)$$

Proof of Lemma 1

We note that in general, we have

$$f(x) \leq Q(x, y) \tag{3.29}$$

for arbitrary $x, y \in \mathbb{R}^p$. In fact, g is convex and differentiable; from Taylor's theorem, we have

$$g(x) = g(y) + (x - y)^T \nabla g(y) + \frac{1}{2}(x - y)^T \nabla^2 g(y + \theta(x - y))(x - y) \leq g(y) + (y - x)^T \nabla g(y) + \frac{L}{2} \|x - y\|_2^2 .$$

Then, the condition that the subderivative of $Q(x, y)$ by x contains 0 is expressed as

$$\nabla g(y) + L(p(y) - y) + \gamma(y) = 0 , \tag{3.30}$$

where $\gamma(y)$ is a subderivative of $h(x)$ when $x = p(y)$. From the convexity of g, h, we can write

$$g(x) \geq g(y) + (x - y)^T \nabla g(y)$$
$$h(x) \geq h(p(y)) + (x - p(y))^T \gamma(y) .$$

From these and (3.30), we have the inequality

$$\begin{aligned}
f(x) - Q(p(y), y) &= g(x) + h(x) - Q(p(y), y) \\
&\geq g(y) + (x - y)^T \nabla g(y) + h(p(y)) + (x - p(y))^T \gamma(y) \\
&\quad - \{g(y) + (p(y) - y)^T \nabla g(y) + \frac{L}{2} \|p(y) - y\|^2 + h(p(y))\} \\
&= -\frac{L}{2} \|p(y) - y\|^2 + (x - p(y))^T (\nabla g(y) + \gamma(y)) \\
&= -\frac{L}{2} \|p(y) - y\|^2 + L(x - y + y - p(y))^T (y - p(y)) \\
&= \frac{L}{2} \|p(y) - y\|^2 + L(x - y)^T (y - p(y)) . \tag{3.31}
\end{aligned}$$

Thus, from (3.29), (3.31), Lemma 1 follows. □

Proof of Proposition 7 If we substitute $x := \beta_*, y := \beta_t$ in Lemma 1 from $p(\beta_t) = \beta_{t+1}$, we have

$$\frac{2}{L}\{f(\beta_*) - f(\beta_{t+1})\} \geq \|\beta_{t+1} - \beta_t\|^2 + 2(\beta_t - \beta_*)^T(\beta_{t+1} - \beta_t)$$

$$= \langle \beta_{t+1} - \beta_t, \beta_{t+1} - \beta_t + 2(\beta_t - \beta_*)\rangle$$

$$= \langle (\beta_{t+1} - \beta_*) - (\beta_t - \beta_*), (\beta_{t+1} - \beta_*) + (\beta_t - \beta_*)\rangle$$

$$= \|\beta_* - \beta_{t+1}\|^2 - \|\beta_* - \beta_t\|^2,$$

where $\langle \cdot, \cdot \rangle$ denotes the inner product in the associated linear space. If we add them over $t = 0, 1, \ldots, k - 1$, we have

$$\frac{2}{L}\{kf(\beta_*) - \sum_{t=0}^{k-1} f(\beta_{t+1})\} \geq \|\beta_* - \beta_k\|^2 - \|\beta_* - \beta_0\|^2. \qquad (3.32)$$

Next, if we substitute $x := \beta_t$, $y := \beta_t$ in Lemma 1, then we have

$$\frac{2}{L}\{f(\beta_t) - f(\beta_{t+1})\} \geq \|\beta_t - \beta_{t+1}\|^2.$$

If we add them after multiplying each by t on both sides over $t = 0, 1, \ldots, k - 1$, we have

$$\frac{2}{L}\sum_{t=0}^{k-1} t\{f(\beta_t) - f(\beta_{t+1})\} = \frac{2}{L}\sum_{t=0}^{k-1}\{tf(\beta_t) - (t+1)f(\beta_{t+1}) + f(\beta_{t+1})\}$$

$$= \frac{2}{L}\{-kf(\beta_k) + \sum_{t=0}^{k-1} f(\beta_{t+1})\} \geq \sum_{t=0}^{k-1} t\|\beta_t - \beta_{t+1}\|^2.$$

$$(3.33)$$

Finally, adding both sides of (3.32) and (3.33), we obtain

$$\frac{2k}{L}\{f(\beta_*) - f(\beta_k)\} \geq \|\beta_* - \beta_k\|^2 - \|\beta_* - \beta_0\|^2 + \sum_{t=0}^{k-1} t\|\beta_t - \beta_{t+1}\|^2$$

$$\geq -\|\beta_0 - \beta_*\|^2,$$

which proves Proposition 7.   □

Exercises 35–46

34. For the function $f(x, y) := \sqrt{x^2 + y^2}$, answer the following questions.

  (a) Find the subderivative of $f(x, y)$ at $(x, y) \neq (0, 0)$.
  (b) Let $p \geq 2$. For $\beta \in \mathbb{R}^p$, Differentiate the L2 norm $\|\beta\|_2$ for $\beta \neq 0$.
  (c) The subderivative of two variables is defined by the set $(u, v) \in \mathbb{R}^2$ such that

$$f(x, y) \geq f(x_0, y_0) + u(x - x_0) + v(y - y_0) \qquad \text{(cf.(3.4))}$$

for $(x, y) \in \mathbb{R}^2$. Find the subderivative of $f(x, y)$ at $(x, y) = (0, 0)$.
Hint Let $x = r \cos \theta$, $y = r \sin \theta$, $u = s \cos \phi$, and $v = s \sin \phi$ ($s \geq 0$, $0 \leq \phi < 2\pi$). Then, $r \geq rs \cos(\theta - \phi)$ is required for all $r \geq 0$, $0 \leq \theta < 2\pi$.

35. Let $X \in \mathbb{R}^{N \times p}$, $y \in \mathbb{R}^N$, and $\lambda > 0$. We wish to find $\beta \in \mathbb{R}^p$ that minimizes

$$\frac{1}{2} \|y - X\beta\|_2^2 + \lambda \|\beta\|_2 . \qquad \text{(cf. (3.5))}$$

(a) Show that the necessary and sufficient condition for $\beta = 0$ to be a solution is $\|X^T y\|_2 \leq \lambda$.
Hint If we take the subderivative, the condition of containing zero is as follows:

$$-X^T(y - X\beta) + \lambda \frac{\beta}{\|\beta\|_2} \ni \begin{bmatrix} 0 \\ 0 \end{bmatrix} \qquad \text{(cf. (3.8))}$$

Substitute $\beta = 0$ into the equation.

(b) Show that if $\beta \neq 0$ has a solution, then β satisfies

$$X^T X\beta = X^T y - \lambda \frac{\beta}{\|\beta\|_2} . \qquad \text{(cf. (3.9))}$$

36. Let $\nu > 0$. Given an initial value of $\beta \in \mathbb{R}^p$, we repeat

$$\gamma := \beta + \nu X^T(y - X\beta) \qquad \text{(cf. (3.10))}$$

$$\beta = \left(1 - \frac{\nu\lambda}{\|\gamma\|_2}\right)_+ \gamma \qquad \text{(cf. (3.11))}$$

to obtain the optimal solution of $\beta \in \mathbb{R}^p$ to Exercise 35, where $(u)_+ := \max\{0, u\}$. As νm, we chose the inverse of the maximum eigenvalue of $X^T X$ and constructed the following procedure. Fill in the blanks and define the function gr. Then, examine the function using the procedure that follows.

```
1   gr = function(X, y, lambda) {
2     nu = 1 / max(2 * eigen(t(X) %*% X)$values)
3     p = ncol(X)
4     beta = rep(1, p); beta.old = rep(0, p)
5     while (max(abs(beta - beta.old)) > 0.001) {
6       beta.old = beta
7       gamma = ## Blank(1) ##
8       beta = ## Blank(2) ##
9     }
10    return(beta)
11  }
12
13  ## Data Generation
```

```
14  n = 100
15  p = 3
16  X = matrix(rnorm(n * p), ncol = p); beta = rnorm(p); epsilon = rnorm(
        n)
17  y = 0.1 * X %*% beta + epsilon
18  ## Display the coefficients that change with lambda
19  lambda = seq(1, 50, 0.5)
20  m = length(lambda)
21  beta = matrix(nrow = m, ncol = p)
22  for (i in 1:m) {
23    est = gr(X, y, lambda[i])
24    for (j in 1:p) beta[i, j] = est[j]
25  }
26  y.max = max(beta); y.min = min(beta)
27  plot(lambda[1]:lambda[m], ylim = c(y.min, y.max),
28       xlab = "lambda", ylab = "Coefficients", type = "n")
29  for (j in 1:p) lines(lambda, beta[, j], col = j + 1)
30  legend("topright", legend = paste("Coefficients", 1:p), lwd = 2, col
        = 2:(p + 1))
31  segments(lambda[1], 0, lambda[m], 0)
```

37. We regard the constant L to be Lipschitz and to satisfy

$$(x - y)^T \nabla^2 g(z)(x - y) \leq L\|x - y\|_2^2 \quad \text{(cf. (3.15))} \tag{3.34}$$

for arbitrary $x, y, z \in \mathbb{R}^p$. Show that when $g(z) = \frac{1}{2}\|y - Xz\|^2$, the Lipschitz constant is at most the maximum eigenvalue of $X^T X$.

38. We modify Exercise 46 (ISTA) as follows to obtain a better performance (FISTA). Let $\{\alpha_t\}$ be the sequence such that $\alpha_1 = 1$ and $\alpha_{t+1} := (1 + \sqrt{1 + 4\alpha_t^2})/2$. We generate $\{\beta\}$ and $\{\gamma\}$ such that $\beta_0 \in \mathbb{R}^p$, $\gamma_1 = \beta_0$, $\beta_t = p(\gamma_t)$

$$\gamma_{t+1} = \beta_t + \frac{\alpha_t - 1}{\alpha_{t+1}}(\beta_t - \beta_{t-1}) \tag{cf. (3.20)}$$

(Iterative Shrinkage-Thresholding Algorithm, ISTA).

(a) Prove $\alpha_t \geq (t + 1)/2$ based on induction.

(b) Fill in the blanks in the program below, replace the function gr in Exercise 36 with $fista$, and examine whether the same execution result is obtained.

```
1   fista = function(X, y, lambda) {
2     nu = 1 / max(2 * eigen(t(X) %*% X)$values)
3     p = ncol(X)
4     alpha = 1
5     beta = rnorm(p); beta.old = rnorm(p)
6     gamma = beta
7     while (max(abs(beta - beta.old)) > 0.001) {
8       print(beta)
9       beta.old = beta
10      w = ## Blank(1) ##
11      beta = ## Blank(2) ##
12      alpha.old = alpha
13      alpha = (1 + sqrt(1 + 4 * alpha ^ 2)) / 2
```

```
14       gamma = beta + (alpha.old - 1) / alpha * (beta - beta.old)
15     }
16     return(beta)
17   }
```

39. To compare the efficiency between the ISTA and FISTA, we construct the following program. Fill in the blanks, and execute the procedure to examine the difference.

```
1  ## Data Generation
2  n = 100; p = 1   # p = 3
3  X = matrix(rnorm(n * p), ncol = p); beta = rnorm(p); epsilon = rnorm(
       n)
4  y = 0.1 * X %*% beta + epsilon
5  lambda = 0.01
6  nu = 1 / max(eigen(t(X) %*% X)$values)
7  p = ncol(X)
8  m = 10
9  ## Performance of ISTA
10 beta = rep(1, p); beta.old = rep(0, p)
11 t = 0; val = matrix(0, m, p)
12 while (t < m) {
13   t = t + 1; val[t, ] = beta
14   beta.old = beta
15   gamma = ## Blank(1) ##
16   beta = ## Blank(2) ##
17 }
18 eval = array(dim = m)
19 val.final = val[m, ]; for (i in 1:m) eval[i] = norm(val[i, ] - val.
       final, "2")
20 plot(1:m, ylim = c(0, eval[1]), type = "n",
21      xlab = "Repetitions", ylab = "L2 Error", main = "Comparison
       bewteen ISTA and FISTA")
22 lines(eval, col = "blue")
23 ## Performance of FISTA
24 beta = rep(1, p); beta.old = rep(0, p)
25 alpha = 1; gamma = beta
26 t = 0; val = matrix(0, m, p)
27 while (t < m) {
28   t = t + 1; val[t, ] = beta
29   beta.old = beta
30   w = ## Blank(3) ##
31   beta = ## Blank(4) ##
32   alpha.old = alpha
33   alpha = ## Blank(5) ##
34   gamma = ## Blank(6) ##
35 }
36 val.final = val[m, ]; for (i in 1:m) eval[i] = norm(val[i, ] - val.
       final, "2")
37 lines(eval, col = "red")
38 legend("topright", c("FISTA", "ISTA"), lwd = 1, col = c("red", "blue"
       ))
```

40. We consider extending the group Lasso procedure for one group to the case of K groups, similar to the procedure for the coordinate decent method, to obtain the solution of

$$\frac{1}{2}\sum_{i=1}^{N}(y_i - \sum_{k=1}^{J} z_{i,k}\theta_k)^2 + \lambda \sum_{k=1}^{K} \|\theta_k\|_2 \quad \text{(cf. (3.1))}. \quad (3.35)$$

Fill in the blanks and construct the group Lasso. Then, execute the function for the following procedure to examine the behavior.

```
1   group.lasso = function(z, y, lambda = 0) {
2     J = length(z)
3     theta = list(); for (j in 1:J) theta[[j]] = rep(0, ncol(z[[j]]))
4     for (m in 1:10) {
5       for (j in 1:J) {
6         r = y; for (k in 1:J) {if (k != j) r = r - ## Blank(1) ##}
7         theta[[j]] = ## Blank(2) ##
8       }
9     }
10    return(theta)
11  }
12
13  ## Data Generation
14  N = 100; J = 2
15  u = rnorm(n); v = u + rnorm(n)
16  s = 0.1 * rnorm(n); t = 0.1 * s + rnorm(n); y = u + v + s + t + rnorm
        (n)
17  z = list(); z[[1]] = cbind(u, v); z[[2]] = cbind(s, t)
18  ## Display of the Coefficients that change with lambda
19  lambda = seq(1, 500, 10); m = length(lambda); beta = matrix(nrow = m,
        ncol = 4)
20  for (i in 1:m) {
21    est = group.lasso(z, y, lambda[i])
22    beta[i, ] = c(est[[1]][1], est[[1]][2], est[[2]][1], est[[2]][2])
23  }
24  y.max = max(beta); y.min = min(beta)
25  plot(lambda[1]:lambda[m], ylim = c(y.min, y.max),
26       xlab = "lambda", ylab = "Coefficient", type = "n")
27  lines(lambda, beta[, 1], lty = 1, col = 2); lines(lambda, beta[, 2],
        lty = 2, col = 2)
28  lines(lambda, beta[, 3], lty = 1, col = 4); lines(lambda, beta[, 4],
        lty = 2, col = 4)
29  legend("topright", legend = c("Group1", "Group1", "Group2", "Group2"
        ),
30         lwd = 1, lty = c(1, 2), col = c(2, 2, 4, 4))
31  segments(lambda[1], 0, lambda[m], 0)
```

41. To contain sparsity not just within a group but also between groups, we extend the formulation (3.35) to

$$\frac{1}{2}\sum_{i=1}^{N}(y_i - \sum_{k=1}^{K} z_{i,k}\theta_k)^2 + \lambda \sum_{k=1}^{K}\{(1-\alpha)\|\theta_k\|_2 + \alpha\|\theta_k\|_1\} \quad (0 < \alpha < 1) \quad \text{(cf. (3.21))} \quad (3.36)$$

(sparse group Lasso). While the active group contains only active variables in the ordinary group Lasso, it may contain nonactive variables in the sparse group Lasso. In other words, the sparse group Lasso extends the group Lasso and allows the active group to contain nonactive variables.

(a) Show that the minimum value except the second term of (3.36) is

$$
\mathcal{S}_{\lambda\alpha}\left(\sum_{i=1}^{N} z_{i,k} r_{i,k}\right) .
$$

(b) Show that the necessary and sufficient condition for $\theta_j = 0$ to be a solution is

$$
\lambda(1-\alpha) \geq \left\| \mathcal{S}_{\lambda\alpha}\left(\sum_{i=1}^{N} z_{i,k} r_{i,k}\right) \right\|_2 .
$$

(c) Show that the update formula for the ordinary group Lasso $\beta \leftarrow \left(1 - \dfrac{\nu\lambda}{\|\gamma\|_2}\right)_{+} \gamma$ becomes

$$
\beta \leftarrow \left(1 - \frac{\nu\lambda(1-\alpha)}{\|\mathcal{S}_{\lambda\alpha}(\gamma)\|_2}\right)_{+} \mathcal{S}_{\lambda\alpha}(\gamma) \qquad\qquad \text{(cf. (3.23))}
$$

in the extended setting.

42. We construct a function that realizes the sparse group Lasso. Fill in the blanks, and execute the procedure in Exercise 40 to examine the validity.

```
1   sparse.group.lasso = function(z, y, lambda = 0, alpha = 0) {
2     J = length(z)
3     theta = list(); for (j in 1:J) theta[[j]] = rep(0, ncol(z[[j]]))
4     for (m in 1:10) {
5       for (j in 1:J) {
6         r = y; for (k in 1:J) {if (k != j) r = r - z[[k]] %*% theta[[k
           ]]}
7         theta[[j]] = ## Blank(1) ##
8       }
9     }
10    return(theta)
11  }
12
13  sparse.gr = function(X, y, lambda, alpha = 0) {
14    nu = 1 / max(2 * eigen(t(X) %*% X)$values)
15    p = ncol(X)
16    beta = rnorm(p); beta.old = rnorm(p)
17    while (max(abs(beta - beta.old)) > 0.001) {
18      beta.old = beta
19      gamma = beta + nu * t(X) %*% (y - X %*% beta)
20      delta = ## Blank(2) ##
21      beta = ## Blank(3) ##
22    }
23    return(beta)
24  }
```

43. We consider the case where some groups overlap. For example, the groups are $\{1, 2, 3\}, \{3, 4, 5\}$ rather than $\{1, 2\}, \{3, 4\}, \{5\}$. We prepare the variables

$\theta_1, \ldots, \theta_K \in \mathbb{R}^p$ such that $\mathbb{R}^p \ni \beta = \sum_{k=1}^{K} \theta_k$ (see Example 31) and find the β value that minimizes

$$\frac{1}{2} \|y - X \sum_{k=1}^{K} \theta_k\|_2^2 + \lambda \sum_{k=1}^{K} \|\theta_k\|_2$$

from the data $X \in \mathbb{R}^{N \times p}$, $y \in \mathbb{R}^N$. We write the first and last three columns of $X \in \mathbb{R}^5$ as $X_1 \in \mathbb{R}^3$ and $X_2 \in \mathbb{R}^3$ and let

$$\theta_1 = \begin{bmatrix} \beta_1 \\ \beta_2 \\ \beta_{3,1} \\ 0 \\ 0 \end{bmatrix}, \quad \theta_2 = \begin{bmatrix} 0 \\ 0 \\ \beta_{3,2} \\ \beta_4 \\ \beta_5 \end{bmatrix} \quad (\beta_3 = \beta_{3,1} + \beta_{3,2}) .$$

Express the conditions equivalent to $\theta_1 = 0$, $\theta_2 = 0$ in terms of θ_1, θ_2, X_1, X_2, y, λ, respectively.

Hint Differentiate the equation to be minimized by $\beta_1, \beta_2, \beta_{3,1}, \beta_{3,2}, \beta_4, \beta_5$.

44. From the observations $X \in \mathbb{R}^{N \times p}$, $\beta \in \mathbb{R}^{p \times K}$, and $y \in \mathbb{R}^{N \times K}$, we find the $\beta \in \mathbb{R}^{p \times K}$ that minimizes

$$L(\beta) := L_0(\beta) + \lambda \sum_{j=1}^{p} \|\beta_j\|_2$$

with

$$L_0(\beta) := \frac{1}{2} \sum_{i=1}^{N} \sum_{k=1}^{K} (y_{i,k} - \sum_{j=1}^{p} x_{i,j} \beta_{j,k})^2 .$$

Show that

$$\hat{\beta}_j = \frac{1}{\sum_{i=1}^{N} x_{i,j}^2} \left(1 - \frac{\lambda}{\| \sum_{i=1}^{N} x_{i,j} r_i^{(j)} \|_2} \right)_+ \sum_{i=1}^{N} x_{i,j} r_i^{(j)}$$

is the solution.

45. From the observations $(x_1, y_1), \ldots, (x_N, y_N) \in \mathbb{R} \times \mathbb{R}$, we regress y_i by $f_1(x_i) + f_2(x_i)$. Let

$$f_1(x; \alpha, \beta) = \alpha + \beta x$$
$$f_2(x; p, q, r) = p \cos x + q \cos 2x + r \cos 3x$$

and $J = 2$, $p_1 = 2$, $p_2 = 3$. To this end, we obtain $\theta_1 = [\alpha, \beta]^T$, $\theta_2 = [p, q, r]^T$ from

$$z_1 = \begin{bmatrix} 1 & x_1 \\ \vdots & \vdots \\ 1 & x_N \end{bmatrix}, \quad z_2 = \begin{bmatrix} \cos x_1 & \cos 2x_1 & \cos 3x_1 \\ \vdots & \vdots & \vdots \\ \cos x_N & \cos 2x_N & \cos 2x_N \end{bmatrix}.$$

Fill in the blanks, and execute the procedure.

```
1   ## Data Generation
2   n = 100; J = 2; x = rnorm(n); y = x + cos(x)
3   z[[1]] = cbind(rep(1, n), x)
4   z[[2]] = cbind(## Blank(1) ##)
5   ## Display the coefficients that change with lambda
6   lambda = seq(1, 200, 5); m = length(lambda); beta = matrix(nrow = m,
        ncol = 5)
7   for (i in 1:m) {
8     est = ## Blank(2) ##
9     beta[i, ] = c(est[[1]][1], est[[1]][2], est[[2]][1], est[[2]][2],
        est[[2]][3])
10  }
11  y.max = max(beta); y.min = min(beta)
12  plot(lambda[1]:lambda[m], ylim = c(y.min, y.max),
13        xlab = "lambda", ylab = "Coefficients", type = "n")
14  lines(lambda, beta[, 1], lty = 1, col = 2); lines(lambda, beta[, 2],
        lty = 2, col = 2)
15  lines(lambda, beta[, 3], lty = 1, col = 4); lines(lambda, beta[, 4],
        lty = 2, col = 4)
16  lines(lambda, beta[, 5], lty = 3, col = 4)
17  legend("topright", legend = c("1", "x", "cos x", "cos 2x", "cos 3x"),
18          lwd = 1, lty = c(1, 2, 1, 2, 3), col = c(2, 2, 4, 4, 4))
19  segments(lambda[1], 0, lambda[m], 0)
20
21  i = 5   # the lambda[5] value is used
22  f.1 = function(x) beta[i, 1] + beta[i, 2] * x
23  f.2 = function(x) beta[i, 3] * cos(x) + beta[i, 4] * cos(2 * x) +
        beta[i, 5] * cos(3 * x)
24  f = function(x) f.1(x) + f.2(x)
25  curve(f.1(x), -5, 5, col = "red", ylab = "Function Value")
26  curve(f.2(x), -5, 5, col = "blue", add = TRUE)
27  curve(f(x), -5, 5, add = TRUE)
28  legend("topleft", legend = c("f = f.1 + f.2", "f.1", "f.2"),
29          col = c(1, "red", "blue"), lwd = 1)
```

46. The following function `gr.multi.lasso` estimates the coefficients of logistic regression with multiple values. Fill in the blanks, and examine whether the procedure runs.

```
1   gr.multi.lasso = function(X, y, lambda) {
2     n = nrow(X); p = ncol(X); K = length(table(y))
3     beta = matrix(1, p, K)
4     gamma = matrix(0, p, K)
5     Y = matrix(0, n, K); for (i in 1:n) Y[i, y[i]] = 1
6     while (norm(beta - gamma, "F") > 10 ^ (-4)) {
7       gamma = beta
8       eta = X %*% beta
9       P = ## Blank(1) ##
10      t = 2 * max(P * (1 - P))
11      R = (Y - P) / t
```

```
12      for (j in 1:p) {
13        r = R + as.matrix(X[, j]) %*% t(beta[j, ])
14        M = ## Blank(2) ##
15        beta[j, ] = sum(X[, j] ^ 2) ^ (-1) *
16          max(1 - lambda / t / sqrt(sum(M ^ 2)), 0) * M
17        R = r - as.matrix(X[, j]) %*% t(beta[j, ])
18      }
19    }
20    return(beta)
21  }
22  ## the procedure to execute
23  df = iris
24  X = cbind(df[[1]], df[[2]], df[[3]], df[[4]])
25  y = c(rep(1, 50), rep(2, 50), rep(3, 50))
26  lambda.seq = c(10, 20, 30, 40, 50, 60, 70, 80, 90, 100, 125, 150)
27  m = length(lambda.seq); p = ncol(X); K = length(table(y))
28  alpha = array(dim = c(m, p, K))
29  for (i in 1:m) {
30    res = gr.multi.lasso(X, y, lambda.seq[i])
31    for (j in 1:p) {for (k in 1:K) alpha[i, j, k] = res[j, k]}
32  }
33  plot(0, xlim = c(0, 150), ylim = c(min(alpha), max(alpha)), type = "n
        ",
34      xlab = "lambda", ylab = "Coefficient", main = "The coefficients
        that change with lambda")
35  for (j in 1:p) {for (k in 1:K) lines(lambda.seq, alpha[, j, k], col =
        j + 1)}
36  legend("topright", legend = c("Sepal Length", "Sepal Width", "Petal
        Length", "Petal Width"),
37        lwd = 2, col = 2:5)
```

Chapter 4
Fused Lasso

Fused Lasso is the problem of finding the $\theta_1, \ldots, \theta_N$ that minimize

$$\frac{1}{2} \sum_{i=1}^{N} (y_i - \theta_i)^2 + \lambda \sum_{i=1}^{N-1} |\theta_i - \theta_{i+1}| \tag{4.1}$$

given observations $y_1, \ldots, y_N$ and a constant $\lambda > 0$

The outputs $\theta_1, \ldots, \theta_N$ are, respectively, close to $y_1, \ldots, y_N$, and θ_i, θ_{i+1} share the same value if y_i, y_{i+1} have close values (Fig. 4.1). Then, for $i = 1, \ldots, N - 1$, the larger the value of $\lambda > 0$, the larger the penalty of $\theta_i \neq \theta_{i+1}$, which means that $\theta_i = \theta_{i+1}$ is more likely; in particular, y_i, y_{i+1} are close. If λ is infinitely large, we have $\theta_1 = \cdots = \theta_N$. We extend the formulation to finding the $\theta_1, \ldots, \theta_N$ that minimize

$$\frac{1}{2} \sum_{i=1}^{N} (y_i - \theta_i)^2 + \mu \sum_{i=1}^{N} |\theta_i| + \lambda \sum_{i=1}^{N-1} |\theta_i - \theta_{i+1}| \tag{4.2}$$

for given $\mu, \lambda \geq 0$ (sparse fused Lasso) . The extended setting penalizes the size of θ_i as well as the difference $\theta_i - \theta_{i+1}$. Another extension is that the observations $y_1, \ldots, y_N$ may be in the $p \geq 1$ dimension in which the difference $\theta_i - \theta_{i+1}$ is evaluated by the (p-dimension) $L2$ norm.

4.1 Applications of Fused Lasso

Before looking at the inside of the fused Lasso procedure, we overview the application of the CRAN package `genlasso` [2] to various data.

© The Author(s), under exclusive license to Springer Nature Singapore Pte Ltd. 2021
J. Suzuki, *Sparse Estimation with Math and R*,
https://doi.org/10.1007/978-981-16-1446-0_4

Fig. 4.1 Fused Lasso. If the dimension is one, adjacent observations y_i are close, and they share the same θ_i, such as $\{y_2, y_3, y_4\}$ and $\{y_5, y_6\}$ above

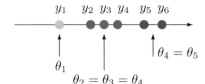

Example 35 We analyze the comparative genomic hybridization (CGH) data via fused Lasso.

https://web.stanford.edu/~hastie/StatLearnSparsity_files/DATA/cgh.html

CGH is a molecular cytogenetic method for analyzing copy number variations (CNVs) relative to the ploidy level in the DNA of a test sample compared to a reference sample without culturing cells. We make the CNVs smooth for $\lambda = 0.1, 1, 10, 100$ (Fig. 4.2).

```
1  library(genlasso)
2  df = read.table("cgh.txt"); y = df[[1]]; N = length(y)
3  out = fusedlasso1d(y)
4  plot(out, lambda = 0.1, xlab = "gene #", ylab = "Copy Number Variation",
5        main = "gene # 1-1000")
```

For the multidimensional case, let $V = \{1, 2, \ldots, N\}$ and E (a subset of $\{\{i, j\} \mid i, j \in V, \ i \neq j\}$) be the vertex and edge sets, respectively, and assume that y_i is at the vertex i. We formulate the fused Lasso problem as minimizing the quantity

$$\frac{1}{2} \sum_{i=1}^{N} (y_i - \theta_i)^2 + \lambda \sum_{\{i,j\}\in E} |\theta_i - \theta_j| \tag{4.3}$$

for a given $\lambda \geq 0$. We consider whether a pair $i, j \in V$ such that $\{i, j\} \in E$ are connected via the fused Lasso procedure. If the edges are connected in one dimension, it is the original fused Lasso (Fig. 4.3).

Example 36 For the novel coronavirus that spread in Japan in 2020, the number of infected people by prefecture up to June 9, 2020 is shown. First, consider the connectivity of fused Lasso when prefectures are adjacent to each other. For $\lambda = 50$, we observe the difference in the number of infected people, but for $\lambda = 150$ or more, only the areas where the infection has spread, such as Tokyo and Hokkaido, are emphasized. In the program, from the adjacency matrix (one if prefectures i, j are concatenated and zero otherwise), the number of columns is 47, and a matrix D with as many rows as the adjacent edges is generated in each line, of which one of the adjacent prefectures has a value of one and the other has a value of -1 (the multidimensional fused Lasso is described in detail in Sect. 4.4).

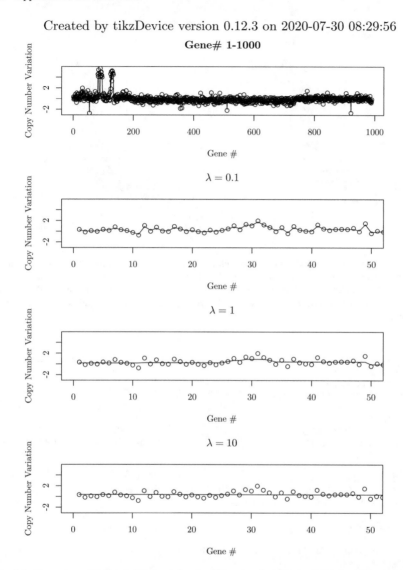

Created by tikzDevice version 0.12.3 on 2020-07-30 08:29:56

Fig. 4.2 Execution of Example 35. Applying fused Lasso, we make the CGH data smooth. For the first graph, we draw the smoothing of the CGH for all genes, and for the three graphs below it, we draw the smoothing for the first fifty genes and $\lambda = 0.1, 1, 10$. We observe that the greater λ is, the more smooth (the less sensitive) the generated line

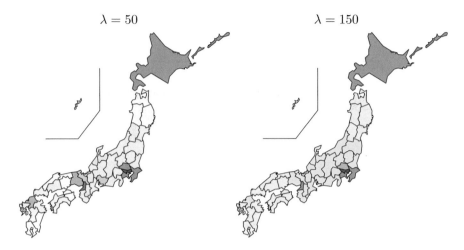

Fig. 4.3 The execution of Example 36. The number of people infected with the novel coronavirus in 2020 until June 9th is displayed in color. When $\lambda = 150$ (right), a difference between a small number of areas can be observed, such as Tokyo and Hokkaido, where the state of emergency declared in April lasted longer than in other prefectures. However, small differences can be observed when $\lambda = 50$ (left)

```
1   library(genlasso)
2   library(NipponMap)
3   mat = read.table("adj.txt")
4   mat = as.matrix(mat)                ## Adjacency matrix of 47 prefectures in
        Japan
5   y = read.table("2020_6_9.txt")
6   y = as.numeric(y[[1]])              ## #infected with corona for each of the 47
        prefectures
7
8   k = 0; u = NULL; v = NULL
9   for (i in 1:46) for (j in (i + 1):47) if (mat[i, j] == 1) {
10    k = k + 1; u = c(u, i); v = c(v, j)
11  }
12  m = length(u)
13  D = matrix(0, m, 47)
14  for (k in 1:m) {D[k, u[k]] = 1; D[k, v[k]] = -1}
15  res = fusedlasso(y, D = D)
16  z = coef(res, lambda = 50)$beta    # lambda = 150
17  cc = round((10 - log(z)) * 2 - 1)
18  cols = NULL
19  for (k in 1:47) cols = c(cols, heat.colors(12)[cc[k]])   ## Colors for each
        of 47 prefectures
20  JapanPrefMap(col = cols, main = "lambda = 50")   ## a function to draw JP map
```

Example 37 Fused Lasso can show not only the difference between the adjacencies of variables but also the second-order difference $\theta_i - 2\theta_{i+1} + \theta_{i+2}$, the third-order difference $\theta_i - 3\theta_{i+1} + 3\theta_{i+2} - \theta_{i+3}$, etc. Differences can be introduced in multiple layers, which is called trend filtering.

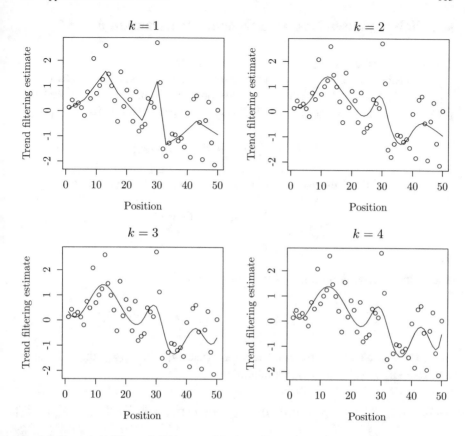

Fig. 4.4 Execution of Example 37 (output of the trend filtering procedure), in which the higher the order k, the more degrees of freedom the curve has, and the easier it is for the curve to follow the data

Using the function `trendfilter` prepared by `genlasso`, we smooth the $\sin \theta$, $0 \leq \theta \leq 2\pi$, data with added noise via a trend filtering of the order $k = 1, 2, 3, 4$. This output is shown in Fig. 4.4.

```
library(genlasso)
n = 50; y = sin(1:n / n * 2 * pi) + rnorm(n)   ## Data Generation
out = trendfilter(y, ord = 3); k = 1  # k = 2, 3, 4
plot(out, lambda = k, main = paste("k = ", k))
```

4.2 Solving Fused Lasso via Dynamic Programming

There are several ways to solve the fused Lasso problem.

We consider the solution based on dynamic programming (N. Johnson, 2013) [14].

To find $\theta_1, \ldots, \theta_N$ that minimize (4.1), we obtain the condition w.r.t. θ_1 such that the optimum $\theta_2, \ldots, \theta_N$ is obtained. In this case, we need to minimize

$$h_1(\theta_1, \theta_2) := \frac{1}{2}(y_1 - \theta_1)^2 + \lambda|\theta_2 - \theta_1|$$

but θ_2 remains. Then, the value of θ_1 when the value of θ_2 is known is

$$\hat{\theta}_1(\theta_2) = \begin{cases} y_1 - \lambda, & y_1 > \theta_2 + \lambda \\ \theta_2, & |y_1 - \theta_2| \leq \lambda \\ y_1 + \lambda, & y_1 < \theta_2 - \lambda \end{cases}.$$

The optimum condition w.r.t. θ_2 is the minimization of

$$\frac{1}{2}(y_1 - \theta_1)^2 + \frac{1}{2}(y_2 - \theta_2)^2 + \lambda|\theta_2 - \theta_1| + \lambda|\theta_3 - \theta_2|.$$

Then, the variables θ_1, θ_3 remain, but if we replace θ_1 with $\hat{\theta}_1(\theta_2)$, the value of θ_2 when we know θ_3, i.e., $\hat{\theta}_2(\theta_3)$ that minimizes

$$h_2(\hat{\theta}_1(\theta_2), \theta_2, \theta_3) := \frac{1}{2}(y_1 - \hat{\theta}_1(\theta_2))^2 + \frac{1}{2}(y_2 - \theta_2)^2 + \lambda|\theta_2 - \hat{\theta}_1(\theta_2)| + \lambda|\theta_3 - \theta_2|$$

can be expressed as a function of θ_3.

Since $\hat{\theta}_1(\theta_2)$ can be expressed by a function $\hat{\theta}_2(\theta_3)$ of θ_3, we write $\hat{\theta}_1(\theta_3)$. If we continue this process, then $\hat{\theta}_1(\theta_N), \ldots, \hat{\theta}_{N-1}(\theta_N)$ are obtained as a function of θ_N. If we substitute them into (4.1), then the problem reduces to the minimization of

$$h_N(\hat{\theta}_1(\theta_N), \ldots, \hat{\theta}_{N-1}(\theta_N), \theta_N, \theta_N)$$
$$:= \frac{1}{2}\sum_{i=1}^{N-1}(y_i - \hat{\theta}_i(\theta_N))^2 + \frac{1}{2}(y_N - \theta_N)^2 + \lambda\sum_{i=1}^{N-2}|\hat{\theta}_i(\theta_N) - \hat{\theta}_{i+1}(\theta_N)| + \lambda|\hat{\theta}_{N-1}(\theta_N) - \theta_N|$$

$$(4.4)$$

w.r.t. θ_N. Suppose that we have successfully obtained the optimum solution $\theta_N = \theta_N^*$ of (4.4). Then, $\theta_{N-1}^* := \hat{\theta}_{N-1}(\theta_N^*)$ is obtained from the value of θ_N^*, $\theta_{N-2}^* = \hat{\theta}_{N-2}(\theta_{N-1}^*)$ is obtained from θ_{N-1}^*, and so on, to obtain the $\theta_1^*, \ldots, \theta_N^*$ that minimize (4.1). We postpone solving the equations w.r.t. $\theta_1, \ldots, \theta_{N-1}$ forward and first solve $\theta_{N-1} = \theta_{N-1}^*, \ldots, \theta_1 = \theta_1^*$ backward after obtaining the solution $\theta_N = \theta_N^*$. This general strategy is called dynamic programming.

It does not seem to be easy to find a function $\hat{\theta}_i(\theta_{i+1})$ except for $i = 1$. However, there exist $L_i, U_i, i = 1, \ldots, N - 1$ such that

$$\hat{\theta}_i(\theta_{i+1}) = \begin{cases} L_i, & \theta_{i+1} < L_i \\ \theta_{i+1}, & L_i \leq \theta_{i+1} \leq U_i \\ U_i, & U_i < \theta_{i+1} \end{cases} .$$

In the actual procedure, we find the optimum value θ_N^* of θ_N and the remaining $\theta_i^* = \hat{\theta}_i(\theta_{i+1}^*)$, $i = N - 1, \ldots, 1$, in a backward manner. For the details, see the Appendix. From the procedure, we have the following proposition.

Proposition 10 (N. Johnson, 2013 [14]) *The algorithm that computes the fused Lasso via dynamic programming completes in $O(N)$.*

Based on the discussion above, we construct a function that computes the fused Lasso.

```
clean = function(z) {
  m = length(z)
  j = 2; while (z[1] >= z[j] && j < m) j = j + 1
  k = m - 1; while (z[m] <= z[k] && k > 1) k = k - 1
  if (j > k) return(z[c(1, m)]) else return(z[c(1, j:k, m)])
}

fused = function(y, lambda = lambda) {
  if (lambda == 0) return(y)
  n = length(y)
  L = array(dim = n - 1)
  U = array(dim = n - 1)
  G = function(i, theta) {
    if (i == 1) theta - y[1]
    else G(i - 1, theta) * (theta > L[i - 1] && theta < U[i - 1]) +
      lambda * (theta >= U[i - 1]) - lambda * (theta <= L[i - 1]) + theta -
      y[i]
  }
  theta = array(dim = n)
  L[1] = y[1] - lambda; U[1] = y[1] + lambda; z = c(L[1], U[1])
  if (n > 2) for (i in 2:(n - 1)) {
    z = c(y[i] - 2 * lambda, z, y[i] + 2 * lambda); z = clean(z)
    m = length(z)
    j = 1; while (G(i, z[j]) + lambda <= 0) j = j + 1
    if (j == 1) {L[i] = z[m]; j = 2}
    else L[i] = z[j - 1] - (z[j] - z[j - 1]) * (G(i, z[j - 1]) + lambda) /
      (-G(i, z[j - 1])+G(i, z[j]))
    k = m; while (G(i, z[k]) - lambda >= 0) k = k - 1
    if (k == m) {U[i] = z[1]; k = m - 1}
    else U[i] = z[k] - (z[k + 1] - z[k]) * (G(i, z[k]) - lambda) /
      (-G(i, z[k]) + G(i, z[k + 1]))
    z = c(L[i], z[j:k], U[i])
  }
  z = c(y[n] - lambda, z, y[n] + lambda); z = clean(z)
  m = length(z)
  j = 1; while (G(n, z[j]) <= 0 && j < m) j = j + 1
  if (j == 1) theta[n] = z[1]
  else theta[n] = z[j - 1] - (z[j] - z[j - 1]) * G(n, z[j - 1]) /(-G(n, z[j
    - 1]) + G(n, z[j]))
```

```
38    for (i in n:2) {
39      if (theta[i] < L[i - 1]) theta[i - 1] = L[i - 1]
40      if (L[i - 1] <= theta[i] && theta[i] <= U[i - 1]) theta[i - 1] = theta[i
        ]
41      if (theta[i] > U[i - 1]) theta[i - 1] = U[i - 1]
42    }
43    return(theta)
44  }
```

We also consider minimizing (4.2) rather than (4.1) (sparse Fused Lasso), which not only smooths the adjacent data but also regularizes $\theta_1, \ldots, \theta_N$ to suppress the sizes.

In reality, we often consider the fused Lasso under $\mu = 0$. Although this is large because we do not care about the size of $\theta = [\theta_1, \ldots, \theta_N]$, we have another reasonable motivation: the solution under the extended setting can be obtained from that under the nonextended setting.

Proposition 11 (Friedman et al., 2007 [10]) *Let $\theta(0)$ be the solution $\theta \in \mathbb{R}^N$ under $\mu = 0$. Then, the solution $\theta = \theta(\mu)$ under $\mu > 0$ is given by $\mathcal{S}_\mu^N(\theta(0))$, where $\mathcal{S}_\mu^N : \mathbb{R}^N \to \mathbb{R}^N$ is such that the $i = 1, \ldots, N$-th element is $\mathcal{S}_\mu(z_i)$ for $z = [z_1, \ldots, z_N]^T$ (see (1.11)).*

For the proof, see the Appendix.

4.3 LARS

To understand the Lasso duality algorithm introduced in the next section, consider the processing of sparse estimation called LARS (least angle regression; B. Efron *et al.*, 2004 [9]) . LARS is a sparse estimation algorithm that performs the same processing as Lasso, but since the amount of calculation is $O(p^2)$ for the number of variables p, Lasso is often used in practice. However, LARS performs processing similar to Lasso, is easy to analyze theoretically and is said to be rich in suggestions.

In linear Lasso, we identify the largest λ such that at least one coefficient is nonzero. Then, such a λ is the largest absolute value λ_0 of $\langle x^{(j)}, y \rangle$ for a variable j, if the squared loss $\|y - X\beta\|^2$ (the first term) is not divided by N, where $x^{(j)}$ is the j-th column of $X \in \mathbb{R}^{N \times p}$, and $y \in \mathbb{R}^N$.

LARS defines a piecewise linear vector $\beta(\lambda)$ (Figure 4.5): suppose we have obtained $\lambda_0, \lambda_1, \lambda_{k-1}, \beta_0 = 0, \beta_1, \beta_{k-1}$, and $\mathcal{S} = \{j_1, \ldots, j_{k-1}\}$ for $k \geq 1$. Then,

1. Define Δ_{k-1} such that the first k and the rest $p - k$ are nonzeros and zeros.
2. Define

$$\beta(\lambda) = \beta_{k-1} + (\lambda_{k-1} - \lambda)\Delta_{k-1} . \tag{4.5}$$

and

$$r(\lambda) = y - X\beta(\lambda) \tag{4.6}$$

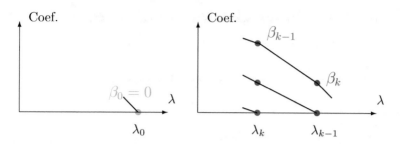

Fig. 4.5 LARS: Δ_{k-1} with k nonzero and $p - k$ zero elements defines the piecewise linear vector $\beta(\lambda) = \beta_{k-1} + (\lambda_{k-1} - \lambda)\Delta_{k-1}$, given $\beta(\lambda_{k-1}) = \beta_{k-1}$

for $\lambda \leq \lambda_{k-1}$.

3. Seek the largest $\lambda_k \leq \lambda_{k-1}$ and $j_k \notin S$ such that that the absolute value of $\langle x^{(j_k)}, r(\lambda_k) \rangle$ is λ_k, and join j_k to the acyive set S.
4. Retrict the ranges of $\beta(\lambda)$ and $r(\lambda)$ to $\lambda_k \leq \lambda$, and define $\beta_k := \beta(\lambda_k)$.

Note that for the third step, if we define $u_j := \langle x_j, r_{k-1} - \lambda_{k-1} X \Delta_{k-1} \rangle$ and $v_j := \langle x_j, X \Delta_{k-1} \rangle$, where $r_{k-1} := y - X \beta_{k-1}$, we choose the largest $t_j := \dfrac{u_j}{v_j \pm 1}$ among $j \notin S$ as λ_k.

If we choose

$$\Delta_{k-1} := \begin{bmatrix} (X_S^T X_S)^{-1} X_S^T r_{k-1}/\lambda_{k-1} \\ 0 \end{bmatrix},$$

then we have $\langle x^{(j_k)}, y \rangle = \lambda$ for $\lambda \leq \lambda_k$, for $k = 0, 1, \ldots, p - 1$, where $X_S \in \mathbb{R}^{N \times k}$ is the sub-matrix of X that consists of $x^{(j)}$ with $j \in S$.

In fact, from (4.6), if we define $H_S := X_S(X_S^T X_S)^{-1} X_S^T$, then we have that

$$r(\lambda) = r_k - \left(1 - \frac{\lambda}{\lambda_{k-1}}\right) H_S r_k = (I - H_S)r_k + \frac{\lambda}{\lambda_{k-1}} H_S r_k .$$

holds. Moreover, from $\langle x^{(j)}, r_k \rangle = \pm \lambda_k$ and $(x^{(j)})^T H_S = (H_S x^{(j)})^T = (x^{(j)})^T$, we have

$$\langle x^{(j)}, r(\lambda) \rangle = \langle x^{(j)}, (I - H_S)r_k + \frac{\lambda}{\lambda_k} H_S r_k \rangle = 0 + \frac{\lambda}{\lambda_k} \langle x^{(j)}, r_k \rangle = \pm \lambda .$$

In other words, in LARS, once one variable joins the active set, they share the same relation $\langle x^{(j)}, r(\lambda) \rangle = \pm \lambda$ until $\lambda = 0$.

For example, in the R language, we can construct the following procedure.

```
lars = function(X, y) {
    X = as.matrix(X); n = nrow(X); p = ncol(X); X.bar = array(dim = p)
    for (j in 1:p) {X.bar[j] = mean(X[, j]); X[, j] = X[, j] - X.bar[j]}
    y.bar = mean(y); y = y - y.bar
    scale = array(dim = p)
```

```
6    for (j in 1:p) {scale[j] = sqrt(sum(X[, j] ^ 2) / n); X[, j] = X[, j] /
       scale[j]}
7    beta = matrix(0, p + 1, p); lambda = rep(0, p + 1)
8    for (i in 1:p) {
9      lam = abs(sum(X[, i] * y))
10     if (lam > lambda[1]) {i.max = i; lambda[1] = lam}
11   }
12   r = y; index = i.max; Delta = rep(0, p)
13   for (k in 2:p) {
14     Delta[index] = solve(t(X[, index]) %*% X[, index]) %*%
15       t(X[, index]) %*% r / lambda[k - 1]
16     u = t(X[, -index]) %*% (r - lambda[k - 1] * X %*% Delta)
17     v = -t(X[, -index]) %*% (X %*% Delta)
18     t = u / (v + 1)
19     for (i in 1:(p - k + 1)) if (t[i] > lambda[k]) {lambda[k] = t[i]; i.max
       = i}
20     t = u / (v - 1)
21     for (i in 1:(p - k + 1)) if (t[i] > lambda[k]) {lambda[k] = t[i]; i.max
       = i}
22     j = setdiff(1:p, index)[i.max]
23     index = c(index, j)
24     beta[k, ] = beta[k - 1, ] + (lambda[k - 1] - lambda[k]) * Delta
25     r = y - X %*% beta[k, ]
26   }
27   for (k in 1:(p + 1)) for (j in 1:p) {beta[k, j] = beta[k, j] / scale[j]}
28   return(list(beta = beta, lambda = lambda))
29 }
```

Example 38 We apply LARS to the U.S. crime data to which we applied Lasso in
Chapter 1. Although the scales of λ are different, they share a similar shape (Fig.
4.6). The figure is displayed via the above function and the following procedure.

```
1    df = read.table("crime.txt"); X = as.matrix(df[, 3:7]); y = df[, 1]
2    res = lars(X, y)
3    beta = res$beta; lambda = res$lambda
4    p = ncol(beta)
5    plot(0:8000, ylim = c(-7.5, 15), type = "n",
6         xlab = "lambda", ylab = "beta", main = "LARS (USA Crime Data)")
7    abline(h = 0)
8    for (j in 1:p) lines(lambda[1:(p)], beta[1:(p), j], col = j)
9    legend("topright",
10          legend = c("Annual Police Funding in \$/Resident", "25 yrs.+ with 4
        yrs. of High School",
11                     "16 to 19 yrs. not in High School ...",
12                     "18 to 24 yrs. in College",
13                     "25 yrs.+ in College"),
14          col = 1:p, lwd = 2, cex = .8)
```

Among the five relations, the first three hold for any probability model. The fourth
is true for the Gaussian distributions from Proposition 16. We only show that the last
relation holds for the Gaussian distribution. The relation if either $k \in A$ or $k \in B$, then
the claim is apparent. If neither $X_A \perp\!\!\!\perp X_k | C$ nor $X_B \perp\!\!\!\perp X_k | C$, then it contradicts
$X_A \perp\!\!\!\perp A_B | X_C$ because the distribution is Gaussian.

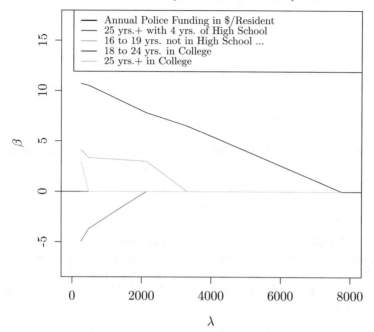

Fig. 4.6 We apply LARS to the U.S. crime data and observe that they share a similar shape

4.4 Dual Lasso Problem and Generalized Lasso

In this section, we solve fused Lasso in a manner different from the method based on dynamic programming. Although the method takes $O(N^2)$ time for N observations, it can solve more general fused Lasso problems. First, we introduce the notion of the dual Lasso problem [29].

Example 39 Given observations $X \in \mathbb{R}^{N \times p}$, $y \in \mathbb{R}^N$, and parameter $\lambda > 0$, we find the $\beta \in \mathbb{R}^p$ that minimizes

$$\frac{1}{2}\|y - X\beta\|_2^2 + \lambda\|\beta\|_1 \tag{4.7}$$

(linear regression Lasso), where the first term is not divided by N (consider that N is contained in λ), which is the minimization of

$$\frac{1}{2}\|r\|_2^2 + \lambda\|\beta\|_1$$

with $r := y - X\beta$. Moreover, if we regard $\alpha \in \mathbb{R}^N$ as a Lagrange coefficient vector, the problem reduces to the minimization of

$$L(\beta, r, \alpha) := \frac{1}{2}\|r\|_2^2 + \lambda\|\beta\|_1 - \alpha^T(r - y + X\beta)$$

w.r.t. β, r. Thus, if we minimize it w.r.t. β, r, we have

$$\min_{\beta \in \mathbb{R}^p}\{-\alpha^T X\beta + \lambda\|\beta\|_1\} = \begin{cases} 0, & \|X^T\alpha\|_\infty \leq \lambda \\ -\infty, & \text{otherwise} \end{cases}$$

$$\min_r \left\{\frac{1}{2}\|r\|_2^2 - \alpha^T r\right\} = -\frac{1}{2}\alpha^T\alpha,$$

where $\|X^T\alpha\|_\infty$ is the maximum of $|x_j^T\alpha|$, in which x_j is the j-th column of X. Therefore, the minimization of (4.7) w.r.t. β, r and the maximization of

$$\frac{1}{2}\{\|y\|_2^2 - \|y - \alpha\|_2^2\} \tag{4.8}$$

over $\alpha \in \mathbb{R}^N$ under $\|X^T\alpha\|_\infty \leq \lambda$, i.e., the minimization of $\frac{1}{2}\|y - \alpha\|_2^2$, are equivalent. In the following, the former and latter are said to be the prime and dual problems.
 This dual problem involves choosing the α value such that $\|X^T\alpha\|_\infty \leq \lambda$ (α surrounded by a polyhedron consisting of p pairs of parallel surfaces) and that minimizes the distance from y.

 Fused Lasso gives an example. For $y_1, \ldots, y_N \in \mathbb{R}$ and $\lambda > 0$, we wish to minimize

$$\frac{1}{2}\|y - \theta\|_2^2 + \lambda\|D\theta\|_1, \tag{4.9}$$

where $D = (D_{i,j}) \in \mathbb{R}^{(N-1)\times N}$ is

$$D_{i,j} = \begin{cases} 1, & j = i \\ -1, & j = i + 1 \\ 0, & \text{otherwise} \end{cases} \tag{4.10}$$

for $i = 1, \ldots, N - 1$.
 The same idea can be applied not just to the one-dimensional fused Lasso but also to trend filtering

$$\frac{1}{2}\sum_{i=1}^{N}(y_i - \theta_i)^2 + \sum_{i=1}^{N-2}|\theta_i - 2\theta_{i+1} + \theta_{i+2}|$$

and its extension

$$\frac{1}{2}\sum_{i=1}^{N}(y_i - \theta_i)^2 + \sum_{i=1}^{N-2}\left|\frac{\theta_{i+2} - \theta_{i+1}}{x_{i+2} - x_{i+1}} - \frac{\theta_{i+1} - \theta_i}{x_{i+1} - x_i}\right| \tag{4.11}$$

by modifying $D = \mathbb{R}^{(N-2)\times N}$ as

$$
D_{i,j} = \begin{cases} 1, & j = i \\ -2, & j = i+1 \\ 1, & j = i+2 \\ 0, & \text{otherwise} \end{cases}, \quad D_{i,j} = \begin{cases} \dfrac{1}{x_{i+1} - x_i}, & j = i \\ -\dfrac{1}{x_{i+1} - x_i} + \dfrac{1}{x_{i+2} - x_{i+1}}, & j = i+1 \\ -\dfrac{1}{x_{i+2} - x_{i+1}}, & j = i+2 \\ 0, & \text{otherwise} \end{cases}.
$$

$$(4.12)$$

Moreover, for the multidimensional case, if the edges in (4.3) consist of m elements $\{i_1, j_1\}, \ldots, \{i_m, j_m\}$, then we can construct the matrix $D \in \mathbb{R}^{m \times N}$ such that $D_{k,i_k} = 1, D_{k,j_k} = -1$ for some $1 \le i_k < i_j \le N$ and the other $N - 2$ are zero for $k = 1, \ldots, m$.

Example 40 We transform the fused Lasso problem into its dual version to solve it. If we let $D \in \mathbb{R}^{m \times p}$ and $\gamma = D\theta$, then (4.9) becomes the minimization of

$$
\frac{1}{2}\|y - \theta\|_2^2 + \lambda\|\gamma\|_1
$$

over $\theta \in \mathbb{R}^p$. If we introduce a Lagrange multiplier α, then we have

$$
\frac{1}{2}\|y - \theta\|_2^2 + \lambda\|\gamma\|_1 + \alpha^T (D\theta - \gamma) .
$$

If we minimize this via θ, γ, we have

$$
\min_\theta \left\{ \frac{1}{2}\|y - \theta\|_2^2 + \alpha^T D\theta \right\} = \frac{1}{2}\|y - D^T\alpha\|_2^2 - \|y\|_2^2 \qquad (4.13)
$$

$$
\min_\gamma \{\lambda\|\gamma\|_1 - \alpha^T\gamma\} = \begin{cases} 0, & \|\alpha\|_\infty \le \lambda \\ -\infty, & \text{otherwise} \end{cases}.
$$

Thus, the dual problem minimizes

$$
\frac{1}{2}\|y - D^T\alpha\|_2^2
$$

under $\|\alpha\|_\infty \le \lambda$. If we obtain the solution $\hat{\alpha}$ of α, then by substituting $\hat{\alpha}$ into the left-hand side of (4.13) to be minimized and differentiating by θ, we have $\hat{\theta} = y - D^T\hat{\alpha}$ and obtain the value of θ.

The prime and dual problems of the generalized Lasso can be extended to the following:

Proposition 12 *For $X \in \mathbb{R}^{N \times p}$, $y \in \mathbb{R}^N$, and $D \in \mathbb{R}^{m \times p}$ ($m \ge 1$), if the prime problem involves finding the $\beta \in \mathbb{R}^p$ that minimizes*

$$
\frac{1}{2}\|y - X\beta\|_2^2 + \lambda\|D\beta\|_1 , \qquad (4.14)
$$

then the dual problem involves finding the $\alpha \in \mathbb{R}^m$ that minimizes

$$\frac{1}{2}\|X(X^TX)^{-1}X^Ty - X(X^TX)^{-1}X^TD^T\alpha\|_2^2 \tag{4.15}$$

under $\|\alpha\|_\infty \leq \lambda$, where we assume that X^TX is nonsingular. If we obtain the solution $\hat{\alpha}$ of α, we obtain β as $\hat{\beta} = y - D^T\hat{\alpha}$.

For the proof, see the Appendix.

We consider solving those problems via the path algorithm (R. Tibshirani and J. Taylor, 2013). The procedure solves the dual problem of the generalized Lasso and makes the CRAN package `genlasso` execute in the first section.

For simplicity, we assume that X in (4.15) is the unit matrix and that DD^T is nonsingular. Then, for each $\lambda \geq 0$, we consider the procedure to find the value of α that minimizes

$$\frac{1}{2}\|y - D^T\alpha\|_2^2 \tag{4.16}$$

under $\|\alpha\|_\infty \leq \lambda$.

To this end, we note the mathematical property below.

Proposition 13 (R. Tibshirani and J. Taylor, 2013) *If $D \in \mathbb{R}^{m \times p}$ satisfies*

$$(DD^T)_{i,i} \geq \sum_{j \neq i} |(DD^T)_{i,j}| \quad (i, j = 1, \ldots, m), \tag{4.17}$$

we have

$$\|\alpha_i(\lambda)\|_\infty = \lambda \implies \|\alpha_i(\lambda')\|_\infty = \lambda', \ \lambda' < \lambda. \tag{4.18}$$

For the proof, consult the paper (R. Tibshirani and J. Taylor, 2011 [29]).

For each λ, each element of the solution $\hat{\alpha}(\lambda)$ is at most λ. The path algorithm uses Proposition 13. The condition (4.17) is met for the ordinary fused Lasso as in (4.10) but does not hold for (4.12). However, the original paper considers a generalization for dealing with the case, and the package `genlasso` implements the idea. This book considers only the case where the matrix D satisfies (4.17).

First, let λ_1 and i_1 be the maximum absolute value among the elements and the element i in the least squares solution $\hat{\alpha}^{(1)}$ of (4.16). We define the sequences $\{\lambda_k\}, \{i_k\}, \{s_k\}$ for $k = 2, \ldots, m$ as follows: For $\mathcal{S} := \{i_1, \ldots, i_{k-1}\}$, let λ_k and $i_k = i$ be the maximum absolute value among the elements and the element $i \notin \mathcal{S}$ in $\alpha_{-\mathcal{S}}$ that minimizes

$$\frac{1}{2}\|y - \lambda D_{\mathcal{S}}^T s - D_{-\mathcal{S}}^T \alpha_{-\mathcal{S}}\|_2^2. \tag{4.19}$$

If $\hat{\alpha}_{i_k} = \lambda_\lambda$, then we have $s_{i_k} = 1$; otherwise, $s_{i_k} = -1$. Here, $D_{\mathcal{S}} \in \mathbb{R}^{k \times p}$, $\alpha_{\mathcal{S}} \in \mathbb{R}^k$ consists of the rows of D, α that correspond to $\mathcal{S}$, and $D_{-\mathcal{S}} \in \mathbb{R}^{(m-k) \times p}$, $\alpha_{-\mathcal{S}} \in \mathbb{R}^{m-k}$ consists of the other rows of D, α.

If we differentiate (4.19) by $\alpha_{-\mathcal{S}}$ to minimize it, we have

$$\alpha_k(\lambda) := \{D_{-\mathcal{S}}D_{-\mathcal{S}}^T\}^{-1}D_{-\mathcal{S}}(y - \lambda D_{\mathcal{S}}^T s) . \tag{4.20}$$

We let $a_i - \lambda b_i$ be the $i \notin \mathcal{S}$-th element of the right-hand side of (4.20), λ_k be the λ such that $a_i - \lambda b_i = \pm\lambda$, and i_k be i. If $a_i - \lambda_k b_i = \lambda_k$, then $s_k = 1$; otherwise, $s_k = -1$. If we let

$$t_i := \frac{a_i}{b_i \pm 1} = \frac{\text{the } i\text{-th element of } \{D_{-\mathcal{S}}D_{-\mathcal{S}}^T\}^{-1}D_{-\mathcal{S}}y}{\text{the } i\text{-th element of}\{D_{-\mathcal{S}}D_{-\mathcal{S}}^T\}^{-1}D_{-\mathcal{S}}D_{\mathcal{S}}^T s \pm 1} ,$$

then the maximum value and its i are λ_k and i_k, respectively, where the sign of $\hat{\alpha}(\lambda_k)_{i_k}$ is s_k. Then, we add i_k to $\mathcal{S}$. For $j \notin \mathcal{S}$, we can compute $\alpha_k(\lambda_k)$ from (4.20). For $j \in \mathcal{S}$, we have $\alpha_k(\lambda_k) = \lambda_k$.

For example, we can construct the following procedure.

```
fused.dual = function(y, D) {
  m = nrow(D)
  lambda = rep(0, m); s = rep(0, m); alpha = matrix(0, m, m)
  alpha[1, ] = solve(D %*% t(D)) %*% D %*% y
  for (j in 1:m) if (abs(alpha[1, j]) > lambda[1]) {
    lambda[1] = abs(alpha[1, j])
    index = j
    if (alpha[1, j] > 0) s[j] = 1 else s[j] = -1
  }
  for (k in 2:m) {
    U = solve(D[-index, ] %*% t(as.matrix(D[-index, , drop = FALSE])))
    V = D[-index, ] %*% t(as.matrix(D[index, , drop = FALSE]))
    u = U %*% D[-index, ] %*% y
    v = U %*% V %*% s[index]
    t = u / (v + 1)
    for (j in 1:(m - k + 1)) if (t[j] > lambda[k]) {lambda[k] = t[j]; h = j;
      r = 1}
    t = u / (v - 1)
    for (j in 1:(m - k + 1)) if (t[j] > lambda[k]) {lambda[k] = t[j]; h = j;
      r = -1}
    alpha[k, index] = lambda[k] * s[index]
    alpha[k, -index] = u - lambda[k] * v
    h = setdiff(1:m, index)[h]
    if (r == 1) s[h] = 1 else s[h] = -1
    index = c(index, h)
  }
  return(list(alpha = alpha, lambda = lambda))
}
```

If we choose the following setting, we have the ordinary (one-dimensional) fused Lasso.

```
m = p - 1; D = matrix(0, m, p); for (i in 1:m) {D[i, i] = 1; D[i, i + 1] =
  -1}
```

Moreover, the solution of the primary problem can be obtained via

$$\hat{\beta}(\lambda) = y - D^T\hat{\alpha}(\lambda)$$

```
1   fused.prime = function(y, D) {
2     res = fused.dual(y, D)
3     return(list(beta = t(y - t(D) %*% t(res$alpha)), lambda = res$lambda))
4   }
```

Example 41 Using the path algorithm introduced above, we draw the graph of the coefficients $\alpha(\lambda)/\hat{\beta}(\lambda)$ of the dual/primary problems with λ in the horizontal axes. (Fig. 4.7).

```
1    p = 8; y = sort(rnorm(p)); m = p - 1; s = 2 * rbinom(m, 1, 0.5) - 1
2    D = matrix(0, m, p); for (i in 1:m) {D[i, i] = s[i]; D[i, i + 1] = -s[i]}
3    par(mfrow = c(1, 2))
4    res = fused.dual(y, D); alpha = res$alpha; lambda = res$lambda
5    lambda.max = max(lambda); m = nrow(alpha)
6    alpha.min = min(alpha); alpha.max = max(alpha)
7    plot(0:lambda.max, xlim = c(0, lambda.max), ylim = c(alpha.min, alpha.max),
         type = "n",
8         xlab = "lambda", ylab = "alpha", main = "Dual Problem")
9    u = c(0, lambda); v = rbind(0, alpha); for (j in 1:m) lines(u, v[, j], col =
         j)
10   res = fused.prime(y, D); beta = res$beta
11   beta.min = min(beta); beta.max = max(beta)
12   plot(0:lambda.max, xlim = c(0, lambda.max), ylim = c(beta.min, beta.max),
         type = "n",
13        xlab = "lambda", ylab = "beta", main = "Prime Problem")
14   w = rbind(0, beta); for (j in 1:p) lines(u, w[, j], col = j)
15   par(mfrow = c(1, 1))
```

We now consider the general problem (4.15) that contains the design matrix $X \in \mathbb{R}^{N \times p}$ with the rank p when the matrix D satisfies (4.17).

If we let $\tilde{y} := X(X^T X)^{-1} X^T y$, $\tilde{D} := D(X^T X)^{-1} X^T$, the problem reduces to minimizing

$$\frac{1}{2} \|\tilde{y} - \tilde{D}\alpha\|_2^2 .$$

For example, we can construct the following R language function.

```
1    fused.dual.general = function(X, y, D) {
2      X.plus = solve(t(X) %*% X) %*% t(X)
3      D.tilde = D %*% X.plus
4      y.tilde = X %*% X.plus %*% y
5      return(fused.dual(y.tilde, D.tilde))
6    }
7    fused.prime.general = function(X, y, D) {
8      X.plus = solve(t(X) %*% X) %*% t(X)
9      D.tilde = D %*% X.plus
10     y.tilde = X %*% X.plus %*% y
11     res = fused.dual.general(X, y, D)
12     m = nrow(D)
13     beta = matrix(0, m, p)
14     for (k in 1:m) beta[k, ] = X.plus %*% (y.tilde - t(D.tilde) %*% res$alpha[
         k, ])
15     return(list(beta = beta, lambda = res$lambda))
16   }
```

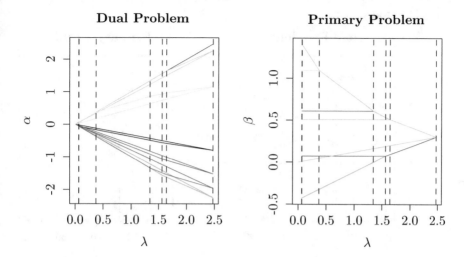

Fig. 4.7 Execution of Example 41: the solution paths of the dual (left) and primary (right) problems w.r.t. $p = 8$ and $m = 7$. We choose the one-dimensional fused Lasso for the matrix D. In both paths, the solution paths merge as λ decreases. The solutions $\alpha \in \mathbb{R}^m$ and $\beta \in \mathbb{R}^p$ of the dual and primary problems are shown by the lines of seven and eight colors, respectively

Example 42 When D is the unit matrix and expresses the one-dimensional fused Lasso (4.10), we show the solution paths of the dual and primary problems (Fig. 4.8). For the unit matrix, we observe the linear Lasso path that we have seen thus far. However, for (4.10), the design matrix is not the unit matrix, and the nature of the path is different. Those executions are due to the following problem.

```
1   n = 20; p = 10; beta = rnorm(p + 1)
2   X = matrix(rnorm(n * p), n, p); y = cbind(1, X) %*% beta + rnorm(n)
3   # D = diag(p)   ## Use one of the two D
4   D = array(dim = c(p - 1, p))
5   for (i in 1:(p - 1)) {D[i, ] = 0; D[i, i] = 1; D[i, i + 1] = -1}
6   par(mfrow = c(1, 2))
7   res = fused.dual.general(X, y, D); alpha = res$alpha; lambda = res$lambda
8   lambda.max = max(lambda); m = nrow(alpha)
9   alpha.min = min(alpha); alpha.max = max(alpha)
10  plot(0:lambda.max, xlim = c(0, lambda.max), ylim = c(alpha.min, alpha.max),
         type = "n",
11       xlab = "lambda", ylab = "alpha", main = "Dual Problem")
12  u = c(0, lambda); v = rbind(0, alpha); for (j in 1:m) lines(u, v[, j], col =
         j)
13  res = fused.prime.general(X, y, D); beta = res$beta
14  beta.min = min(beta); beta.max = max(beta)
15  plot(0:lambda.max, xlim = c(0, lambda.max), ylim = c(beta.min, beta.max),
         type = "n",
16       xlab = "lambda", ylab = "beta", main = "Primary Problem")
17  w = rbind(0, beta); for (j in 1:p) lines(u, w[, j], col = j)
18  par(mfrow = c(1, 1))
```

In general, suppose that we minimize a function as

(a) D : Unit matrix

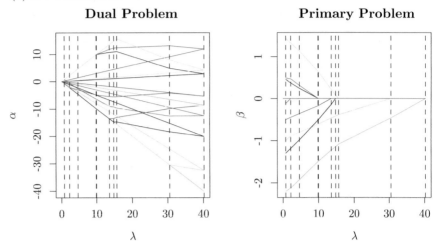

(b) D : One-dimensional fused Lasso

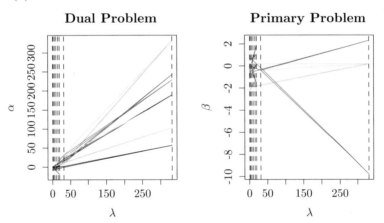

Fig. 4.8 Execution of Example 42. The solution paths of the dual and primary paths when the design matrix is not the unit matrix. **a** When D is the unit matrix, the problem reduces to the linear Lasso. **b** When D expresses the one-dimensional fused Lasso, a completely different shape appears in the solution path

$$g(\beta_1, \ldots, \beta_p) + \lambda h(\beta_1, \ldots, \beta_p)$$

for convex g, h with g being differentiable. Then, if the second term is separated as convex p one-variable functions such as $h(\beta_1, \ldots, \beta_p) = \sum_{j=1}^{p} h_j(\beta_j)$, it is known that the update obtained by differentiating the objective function by β converges to the optimum value (Tseng, 1988).

However, the primary problem of fused Lasso does not satisfy separability, and we cannot apply the coordinate descent method. Thus, we consider the dynamic programming method and the dual problems. We consider another way to solve the fused Lasso problem.

4.5 ADMM

This section considers a different approach that is applicable to general problems to solve the fused Lasso.

Let $A \in \mathbb{R}^{d \times m}$, $B \in \mathbb{R}^{d \times n}$, and $c \in \mathbb{R}^d$. We assume that $f : \mathbb{R}^m \to \mathbb{R}$ and $g : \mathbb{R}^n \to \mathbb{R}$ are convex and that f is differentiable. We formulate the problem of finding $\alpha \in \mathbb{R}^m$, $\beta \in \mathbb{R}^n$ such that

$$A\alpha + B\beta = c \qquad (4.21)$$

$$f(\alpha) + g(\beta) \text{ that minimizes} \qquad (4.22)$$

via the Lagrange multiplier

$$L(\alpha, \beta, \gamma) := f(\alpha) + g(\beta) + \gamma^T(A\alpha + B\beta - c) \to \min \quad (\gamma \in \mathbb{R}^d : \text{undeterminedconstant}) .$$

In general, we have

$$\inf_{\alpha, \beta} L(\alpha, \beta, \gamma) \le L(\alpha, \beta, \gamma) \le \sup_{\gamma} L(\alpha, \beta, \gamma) .$$

Then, the minimum and maximum values of the primary and dual problems coincide, respectively. Although in general, the equality in

$$\sup_{\gamma} \inf_{\alpha, \beta} L(\alpha, \beta, \gamma) \le \inf_{\alpha, \beta} \sup_{\gamma} L(\alpha, \beta, \gamma)$$

may or may not hold, in this case, it is known that if the objective function and the constraints are positive, the equality holds (Slater condition) .

In this book, we introduce a constant $\rho > 0$ to define the extended Lagrangian

$$L_\rho(\alpha, \beta, \gamma) = f(\alpha) + g(\beta) + \gamma^T(A\alpha + B\beta - c) + \frac{\rho}{2}\|A\alpha + B\beta - c\|^2 . \quad (4.23)$$

Setting the initial values $\alpha_0 \in \mathbb{R}^m$, $\beta_0 \in \mathbb{R}^n$, $\gamma_0 \in \mathbb{R}^d$, we update the following steps for $t = 1, 2, \ldots$ (ADMM, Alternating Direction Method of Multipliers):

1. let α_{t+1} be the α that minimizes $L_\rho(\alpha, \beta_t, \gamma_t)$
2. let β_{t+1} be the β that minimizes $L_\rho(\alpha_{t+1}, \beta, \gamma_t)$
3. $\gamma_{t+1} \leftarrow \gamma_t + \rho(A\alpha_{t+1} + B\beta_{t+1} - c)$

We show that the solution of (4.21), (4.22) can be obtained by repeating the above three steps. The ADMM is a general method used to solve convex optimization problems, not just for fused Lasso, and in the following chapters, we continue to apply the ADMM to other classes of sparse estimation problems.

Example 43 If we apply as a generalized Lasso

$$L_\rho(\alpha, \beta, \gamma) := \frac{1}{2}\|y - \alpha\|_2^2 + \lambda\|\beta\|_1 + \mu^T(D\alpha - \beta) + \frac{\rho}{2}\|D\alpha - \beta\|^2$$

then we have

$$\frac{\partial L_\rho}{\partial \alpha} = \alpha - y + D^T\gamma_t + \rho D^T(D\alpha - \beta_t)$$

$$\frac{\partial L_\rho}{\partial \beta} = -\gamma_t + \rho(\beta - D\alpha_{t+1}) + \lambda \begin{cases} 1, & \beta > 0 \\ [-1, 1], & \beta = 0 \\ -1, & \beta < 0 \end{cases},$$

which means that the update formula is as follows:

$$\begin{cases} \alpha_{t+1} \leftarrow (I + \rho D^T D)^{-1}(y + D^T(\rho\beta_t - \gamma_t)) \\ \beta_{t+1} \leftarrow \mathcal{S}_\lambda(\rho D\alpha_{t+1} + \gamma_t)/\rho \\ \gamma_{t+1} \leftarrow \gamma_t + \rho(D\alpha_{t+1} - \beta_{t+1}) \end{cases},$$

where $A \in \mathbb{R}^{d \times m}$, $B \in \mathbb{R}^{d \times n}$, $c \in \mathbb{R}^d$, $f : \mathbb{R}^m \to \mathbb{R}$, and $g : \mathbb{R}^n \to \mathbb{R}$ are, respectively, $A = D$, $B = -I$, $c = 0$, $f = \frac{1}{2}\|y - \alpha\|^2$, and $g = \|\beta\|_1$.

To realize the generalized Lasso, we can, for example, construct the following function.

```
1   admm = function(y, D, lambda) {
2     K = ncol(D); L = nrow(D)
3     theta.old = rnorm(K); theta = rnorm(K); gamma = rnorm(L); mu = rnorm(L)
4     rho = 1
5     while (max(abs(theta - theta.old) / theta.old) > 0.001) {
6       theta.old = theta
7       theta = solve(diag(K) + rho * t(D) %*% D) %*% (y + t(D) %*% (rho * gamma
          - mu))
8       gamma = soft.th(lambda, rho * D %*% theta + mu) / rho
9       mu = mu + rho * (D %*% theta - gamma)
10    }
11    return(theta)
12  }
```

Fig. 4.9 Execution of
Example 44. We obtain the
solution path of the CGH
data in Example 35

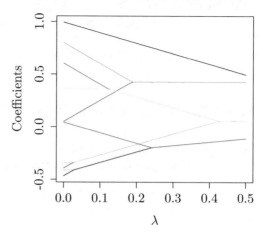

Example 44 We apply the CGH data in Example 35 to the ADMM to obtain the
solution path of the primary problem (Fig. 4.9). For the ADMM, unlike the path
algorithm in the previous section, we obtain $\theta \in \mathbb{R}^N$ for a specific $\lambda \geq 0$.

```
1  df = read.table("cgh.txt"); y = df[[1]][101:110]; N = length(y)
2  D = array(dim = c(N - 1, N)); for (i in 1:(N - 1)) {D[i, ] = 0; D[i, i] = 1;
       D[i, i + 1] = -1}
3  lambda.seq = seq(0, 0.5, 0.01); M = length(lambda.seq)
4  theta = list(); for (k in 1:M) theta[[k]] = admm(y, D, lambda.seq[k])
5  x.min = min(lambda.seq); x.max = max(lambda.seq)
6  y.min = min(theta[[1]]); y.max = max(theta[[1]])
7  plot(lambda.seq, xlim = c(x.min, x.max), ylim = c(y.min, y.max), type = "n",
8        xlab = "lambda", ylab = "Coefficients", main = "Fused Lasso Solution
       Path")
9  for (k in 1:N) {
10     value = NULL; for (j in 1:M) value = c(value, theta[[j]][k])
11     lines(lambda.seq, value, col = k)
12  }
```

It is known that the ADMM satisfies the following theoretical properties.

Proposition 14 Suppose that the functions f, g are convex, that f is differentiable,
and that the ranks of A, B are m, n. If the optimum solution (α_*, β_*) of (4.21), (4.22)
exists, the sequences $\{\alpha_t\}, \{\beta_t\}, \{\gamma_t\}$ satisfy the following properties:

1. The sequence $\{p_t\}$ with $p_t := f(\alpha_t) + g(\beta_t)$ converges to the optimum p_*.
2. The sequences $\{\alpha_t\}, \{\beta_t\}$ respectively converge to the unique values α_*, β_*.
3. The sequence $\{\gamma_t\}$ converges to a unique γ_* that is the dual solution of (4.21),
   (4.22).

For the proof of Proposition 14, see the paper by Joao F. C. Mota, 2011 [21]. Similar
proofs can be found in the following: S. Boyd et al., *Distributed optimization and*

statistical learning via the alternating method of multipliers (2011) [5] does not satisfy the full rank of A, B and does not prove the second property. D. Bertsekas et al., *Convex analysis and optimization* (2003) [4] proves the second property, assuming that B is the unit matrix. They do not contain anything complicated, although the derivation is lengthy. In particular, S. Boyd (2011) [5] can be easily understood.

Appendix: Proof of Proposition

Proposition 10 (N. Johnson, 2013 [14]) *The algorithm that computes the fused Lasso via dynamic programming completes in $O(N)$.*

Proof   In the following, for $i = 1, \ldots, N - 1$, we consider how to obtain $\theta_i = \theta_i^*$ when we know $\theta_{i+1} = \theta_{i+1}^*$ is optimum. For the quantity

$$\frac{1}{2} \sum_{j=1}^{i-1} \{y_j - \hat{\theta}_j(\theta_i)\}^2 + \frac{1}{2}(y_i - \theta_i)^2 + \lambda \sum_{j=1}^{i-2} |\hat{\theta}_j(\theta_i) - \hat{\theta}_{j+1}(\theta_i)| + \lambda |\hat{\theta}_{i-1}(\theta_i) - \theta_i|$$

that is obtained by removing the last term $\lambda |\theta_i - \theta_{i+1}^*|$ in

$$h_i(\hat{\theta}_1(\theta_i), \ldots, \hat{\theta}_{i-1}(\theta_i), \theta_i, \theta_{i+1}^*)$$
$$= \frac{1}{2} \sum_{j=1}^{i-1} \{y_j - \hat{\theta}_j(\theta_i)\}^2 + \frac{1}{2}(y_i - \theta_i)^2 + \lambda \sum_{j=1}^{i-2} |\hat{\theta}_j(\theta_i) - \hat{\theta}_{j+1}(\theta_i)| + \lambda |\hat{\theta}_{i-1}(\theta_i) - \theta_i| + \lambda |\theta_i - \theta_{i+1}^*|$$

we differentiate by θ_i to define

$$g_i(\theta_i) := - \sum_{j=1}^{i-1} \{y_j - \hat{\theta}_j(\theta_i)\} \frac{d\hat{\theta}_j(\theta_i)}{d\theta_i} - (y_i - \theta_i) + \lambda \sum_{j=1}^{i-2} \frac{d}{d\theta_i} |\hat{\theta}_j(\theta_i) - \hat{\theta}_{j+1}(\theta_i)| + \lambda \frac{d}{d\theta_i} |\hat{\theta}_{i-1}(\theta_i) - \theta_i|.$$

As we notice later, we will have either $\hat{\theta}_j(\theta) = \theta + a$ or $\hat{\theta}_j(\theta) = b$ for the constants a, b, which means that $\dfrac{d\hat{\theta}_j(\theta_i)}{d\theta_i}$ is either zero or one for $j = 1, \ldots, i - 1$. For the other terms that contain absolute values, we cannot differentiate and apply the subderivative.

Thus, for each of the cases, we can obtain $\theta_i^* = \hat{\theta}_i(\theta_{i+1}^*)$:

1. For $\theta_i > \theta_{i+1}^*$, if we differentiate $\lambda |\theta_i - \theta_{i+1}^*|$ by θ_i, it becomes λ. Then, the θ_i value does not depend on θ_{i+1}^*. We let the solution θ_i of

$$\frac{d}{d\theta_i} h_i(\hat{\theta}_1(\theta_i), \ldots, \hat{\theta}_{i-1}(\theta_i), \theta_i, \theta_{i+1}^*) = g_i(\theta_i) + \lambda = 0$$

be $\hat{\theta}_i(\theta_{i+1}^*) := L_i$.

2. For $\theta_i < \theta^*_{i+1}$, if we differentiate $\lambda|\theta_i - \theta^*_{i+1}|$ by θ_i, it becomes $-\lambda$. Then, the θ_i value dose not depend on θ^*_{i+1}. We let the solution of θ_i

$$\frac{d}{d\theta_i} h_i(\hat{\theta}_1(\theta_i), \ldots, \hat{\theta}_{i-1}(\theta_i), \theta_i, \theta^*_{i+1}) = g_i(\theta_i) - \lambda = 0$$

be $\hat{\theta}_i(\theta^*_{i+1}) := U_i$.

3. For the other cases, i.e., for $L_i < \theta^*_{i+1} < U_i$, we have $\hat{\theta}_i(\theta^*_{i+1}) = \theta^*_{i+1}$.

For example, if $i = 1$, then we have $g_1(\theta_1) = \theta_1 - y_1$. If $\theta_1 > \theta^*_2$, then solving $g_1(\theta_1) + \lambda = 0$, we have $\hat{\theta}_1(\theta^*_2) = y_1 - \lambda = L_1$. If $\theta_1 < \theta^*_2$, then solving $g_1(\theta_1) - \lambda = 0$, we have $\hat{\theta}_1(\theta^*_2) = y_1 + \lambda = U_1$. Furthermore, if $L_1 < \theta^*_2 < U_1$, we have $\hat{\theta}_1(\theta^*_2) = \theta^*_2$. Thus, we have

$$L_1 = \hat{\theta}_1(\theta^*_2) > \theta^*_2 \implies \lambda|\theta^*_2 - \hat{\theta}_1(\theta^*_2)| = -\lambda(\theta^*_2 - L_1)$$
$$U_1 = \hat{\theta}_1(\theta^*_2) < \theta^*_2 \implies \lambda|\theta^*_2 - \hat{\theta}_1(\theta^*_2)| = \lambda(\theta^*_2 - U_1)$$
$$L_1 \leq \theta^*_2 \leq U_1 \iff \hat{\theta}_1(\theta^*_2) = \theta^*_2 ,$$

which means that

$$\hat{\theta}_1(\theta_2) := \begin{cases} L_1 = y_1 - \lambda, & \theta_2 < L_1 \\ \theta_2, & L_1 \leq \theta_2 \leq L_2 \\ U_1 = y_1 + \lambda, & U_1 < \theta_2 \end{cases} .$$

Therefore, we can write $g_2(\theta_2)$ as follows:

$$g_2(\theta_2) = \frac{d}{d\theta_2} \left\{ \frac{1}{2}(y_1 - \hat{\theta}(\theta_2))^2 + \frac{1}{2}(y_2 - \theta_2)^2 + \lambda|\hat{\theta}_1(\theta_2) - \theta_2| \right\}$$
$$= \begin{cases} \theta_2 - y_2 - \lambda, & \theta_2 < L_1 \\ 2\theta_2 - y_2 - y_1, & L_1 \leq \theta_2 \leq U_1 \\ \theta_2 - y_2 + \lambda, & U_1 < \theta_2 \end{cases}$$
$$= g_1(\theta_2)I[L_1 \leq \theta_2 \leq U_1] + \lambda I[\theta_2 > U_1] - \lambda I[\theta_2 < L_1] + \theta_2 - y_2 ,$$

where if the condition A is true, then $I[A] = 1$; otherwise, $I[A] = 0$. Similarly, we solve

$$0 = \frac{d}{d\theta_2} h_2(\hat{\theta}_1(\theta_2), \theta_2, \theta^*_3) = \begin{cases} g_2(\theta_2) + \lambda, & \theta_2 > \theta^*_3 \\ g_2(\theta_2) - \lambda, & \theta_2 < \theta^*_3 \end{cases}$$

and we obtain L_2, U_2. Moreover, for $i = 2, \ldots, N$, we have

$$g_i(\theta_i) = g_{i-1}(\theta_i)I[L_{i-1} \leq \theta_i \leq U_{i-1}] + \lambda I[\theta_i > U_{i-1}] \qquad (4.24)$$
$$- \lambda I[\theta_i < L_{i-1}] + \theta_i - y_i .$$

On the other hand, $g_i(\theta_i)$ is a line that has a piecewise nonnegative slope, and the knots are

$$L_1, \ldots, L_{i-1}, U_1, \ldots, U_{i-1} .$$

Then, the first three terms of $g_i(\theta_i)$ are respectively between $-\lambda$ and λ, and the solution of $g_i(\theta_i) \pm \lambda = 0$ ranges over $y_i - 2\lambda \le \theta_i \le y_i + 2\lambda$. In fact, from (4.24), we have

$$g_i(\theta_i) \begin{cases} > \lambda, & \theta_i > y_i + 2\lambda \\ < -\lambda, & \theta_i < y_i - 2\lambda \end{cases} .$$

There are at most $2i$ knots containing the two, which we express as $x_1 < \cdots < x_{2i}$. If $g_i(x_k) + \lambda \le 0$ and $g_i(x_{k+1}) + \lambda \ge 0$, then

$$L_i := x_k + (x_{k+1} - x_k) \frac{|g_i(x_k) + \lambda|}{|g_i(x_k) + \lambda| + |g_i(x_{k+1}) + \lambda|}$$

is the solution. Similarly, we let U_i be the θ_i such that $g_i(\theta_i) - \lambda = 0$. In particular, for $i = N$, we have $\theta_N = \theta_{N+1}$, and no last term exists as in $h_2, \ldots, h_{N-1}$. Thus, we obtain θ_N such that

$$0 = \frac{d}{d\theta_N} h_N(\hat{\theta}_1(\theta_N), \ldots, \hat{\theta}_{N-1}(\theta_N), \theta_N, \theta_N) = g_N(\theta_N) = 0$$

rather than $g_N(\theta_N) \pm \lambda = 0$.

Finally, we evaluate the efficiency of the procedure. If

$$x_1 < \cdots < x_r \le L_{i-1} < x_{r+1} < \cdots < x_{s-1} < U_{i-1} \le x_s < \cdots < x_{2i} ,$$

we may exclude $x_1, \ldots, x_r$ and $x_s, \ldots, x_{2i}$ from the search of L_i, U_i. In fact, from (4.24), we find that the solution of $g_i(\theta_i) \pm \lambda = 0$ does not depend on U_j above U_{i-1} and L_j below L_{i-1}, $j = 1, \ldots, i - 2$, which means that g_i is constructed without them and that $y_i - 2\lambda$, L_{i-1}, $x_{r+1}, \ldots, x_{s-1}$, U_{i-1}, $y_i + 2\lambda$ are the knots.

Moreover, the knots used to search for L_i, U_i are used to search for L_j, U_j ($j = i + 1, \ldots, N - 1$). However, if we remove a knot, then it will not be used for future searches. In addition, at most four knots are added each time, which means that at most $4N$ knots are added in total. Thus, the number of knots that are excluded is at most $4N$ in total as well if we search for L_i, U_i from the outside. Hence, the computation is proportional to at most N.

Finally, if we apply the following step

$$\theta_i^* := \begin{cases} U_i, & \theta_{i+1}^* > U_i \\ \theta_{i+1}^*, & L_i \le \theta_{i+1}^* \le U_i \\ L_i, & \theta_{i+1}^* < L_i \end{cases}$$

for $i = N - 1, \ldots, 1$, we obtain $\{\theta_i^*\}_{i=1}^N$. $\square$

Proposition 11 (Friedman et al., 2007 [10]) *Let $\theta(0)$ be the solution $\theta \in \mathbb{R}^N$ under $\mu = 0$. Then, the solution $\theta = \theta(\mu)$ under $\mu > 0$ is given by $S_\mu^N(\theta(0))$, where $S_\mu^N : \mathbb{R}^N \to \mathbb{R}^N$ is such that the $i = 1, \ldots, N$-th element is $S_\mu(z_i)$ for $z = [z_1, \ldots, z_N]^T$ (see (1.11)).*

Proof   Differentiating (4.2) by θ_i ($i = 2, \ldots, N-1$) and equating it to zero, we have

$$-y_i + \theta_i(\mu) + \mu \partial |\theta_i| + \lambda \partial |\theta_i - \theta_{i-1}| + \lambda \partial |\theta_i - \theta_{i+1}| = 0 \quad (i = 2, \ldots, N-1) .$$

Thus, it is sufficient to show that for $\mu \geq 0$ and each $i = 2, \ldots, N-1$, there exist

$$s_i(\mu) = \begin{cases} 1, & \theta_i > 0 \\ [-1, 1], & \theta_i = 0 \\ -1, & \theta_i < 0 \end{cases}, \quad t_i(\mu) = \begin{cases} 1, & \theta_i > \theta_{i-1} \\ [-1, 1], & \theta_i = \theta_{i-1} \\ -1, & \theta_i < \theta_{i-1} \end{cases}, \quad u_i(\mu) = \begin{cases} 1, & \theta_i > \theta_{i+1} \\ [-1, 1], & \theta_i = \theta_{i+1} \\ -1, & \theta_i < \theta_{i+1} \end{cases}$$

such that

$$f_i(\mu) := -y_i + \theta_i(\mu) + \mu s_i(\mu) + \lambda t_i(\mu) + \lambda u_i(\mu) = 0 \tag{4.25}$$
$$\theta_i(\mu) = S_\mu(\theta_i(0)) , \tag{4.26}$$

where we write $\theta_i(\mu)$ simply as θ_i.

Notably, we may assume that $(s_i(0), t_i(0), u_i(0))$ exists such that $f_i(0) = 0$. In fact, $\theta_i(0)$ satisfies $f_i(0) = 0$, and

$$f_i(0) = -y_i + \theta_i(0) + \lambda t_i(0) + \lambda u_i(0) = 0 . \tag{4.27}$$

Moreover, since from the definition of (4.26), we observe that $S_\mu(x)$ monotonically decreases with $x \in \mathbb{R}$, we have

$$\begin{cases} \theta_i(0) = \theta_j(0) \implies \theta_i(\mu) = S_\mu(\theta_i(0)) = S_\mu(\theta_j(0)) = \theta_j(\mu) \\ \theta_i(0) < \theta_j(0) \implies \theta_i(\mu) = S_\mu(\theta_i(0)) \leq S_\mu(\theta_j(0)) = \theta_j(\mu) \end{cases}$$

for $\mu \geq 0$ and each $i, j = 1, \ldots, N$ ($i \neq j$). Hence,

$$t_i(0) = \begin{cases} 1, & \theta_i(0) > \theta_{i-1}(0) \\ [-1, 1], & \theta_i(0) = \theta_{i-1}(0) \\ -1, & \theta_i(0) < \theta_{i-1}(0) \end{cases}$$

$$\implies t_i(0) = \begin{cases} 1 \text{ or } [-1.1], & \theta_i(0) > \theta_{i-1}(0) \\ [-1, 1], & \theta_i(0) = \theta_{i-1}(0) \\ -1 \text{ or } [-1, 1], & \theta_i(0) < \theta_{i-1}(0) \end{cases}$$

$$\implies t_i(\mu) = [-1, 1] \quad .$$

Hence, if we regard $t_i(0)$, $t_i(\mu)$ as a set similar to $\{1\}$, $[-1, 1]$, $\{-1\}$, we have

$$t_i(0) \subseteq t_i(\mu) . \tag{4.28}$$

Similarly, we have

$$u_i(0) \subseteq u_i(\mu) . \tag{4.29}$$

From (4.26), we have $\theta_i(\mu) = \theta_i(0) \pm \mu$ and $\theta_i(\mu) = 0$ for the i such that $|\theta_i(0)| > \mu$ and for the i such that $|\theta_i(0)| \leq \mu$, respectively.

For the first case, we have

$$\begin{cases} \theta_i(0) > \mu \iff \theta_i(\mu) = \theta_i(0) - \mu \iff s_i(0) = 1 \\ \theta_i(0) < -\mu \iff \theta_i(\mu) = \theta_i(0) + \mu \iff s_i(0) = -1 \end{cases} .$$

Moreover, from $s_i(0) \subseteq s_i(\mu)$, we may assume $s_i(\mu) = s_i(0)$. From (4.28), (4.29), if we substitute the $t_i(0), u_i(0)$ values into $t_i(\mu), u_i(\mu)$, the relation holds. Hence, substituting $\theta_i(\mu) = \theta_i(0) - s_i(0)\mu$ and $(s_i(\mu), t_i(\mu), u_i(\mu)) = (s_i(0), t_i(0), u_i(0))$ into (4.25), from (4.25), (4.27), we have

$$f_i(\mu) = -y_i + \theta_i(0) - s_i(0)\mu + s_i(0)\mu + \lambda t_i(0) + \lambda u_i(0) = 0 .$$

For the latter case, the subderivative of $s_i(0)$ should be $[-1, 1]$ for the i such that $|\theta_i(0)| \leq \mu$. Thus, if we input $s_i(\mu) = \theta_i(0)/\mu \in [-1, 1]$, $s_i(\mu)$ becomes a subdifferential of $|\theta_i(\mu)|$. Also from (4.28), (4.29), the relation holds even if we substitute the $t_i(0), u_i(0)$ values into $t_i(\mu), u_i(\mu)$. Thus, substituting $\theta_i(\mu) = 0$ and $(s_i(\mu), t_i(\mu), u_i(\mu)) = (\theta_i(0)/\mu, t_i(0), u_i(0)) f_i(\mu) = 0$, from (4.25), (4.27), we have

$$f_i(\mu) = -y_i + \mu \cdot \frac{\theta_i(0)}{\mu} + \lambda t_i(0) + \lambda u_i(0) = 0 .$$

Proposition 12 *For $X \in \mathbb{R}^{N \times p}$, $y \in \mathbb{R}^N$, and $D \in \mathbb{R}^{m \times p}$ ($m \geq 1$), if the prime problem finds the $\beta \in \mathbb{R}^p$ that minimizes*

$$\frac{1}{2}\|y - X\beta\|_2^2 + \lambda\|D\beta\|_1 , \tag{4.14}$$

then the dual problem finds the $\alpha \in \mathbb{R}^m$ that minimizes

$$\frac{1}{2}\|X(X^T X)^{-1} X^T y - X(X^T X)^{-1} X^T D^T \alpha\|_2^2 \tag{4.15}$$

under $\|\alpha\|_\infty \leq \lambda$, where we assume that $X^T X$ is nonsingular. If we obtain the solution $\hat{\alpha}$ of α, we obtain β as $\hat{\beta} = y - D^T \hat{\alpha}$.

Proof: If we let $D \in \mathbb{R}^{m \times p}$, $\gamma = D\beta$, then the primary problem can be regarded as the minimization of

$$\frac{1}{2}\|y - X\beta\|_2^2 + \lambda\|\gamma\|_1$$

w.r.t. $\beta \in \mathbb{R}^p$. Introducing the Lagrange multiplier α, we have

$$\frac{1}{2}\|y - X\beta\|_2^2 + \lambda\|\gamma\|_1 + \alpha^T(D\beta - \gamma)\cdot$$

and if we take the minimization of this over β, γ, then we have

$$\min_\beta\left\{\frac{1}{2}\|y - X\beta\|_2^2 + \alpha^T D\beta\right\} = \frac{1}{2}\|y - X(X^TX)^{-1}(X^Ty - D^T\alpha)\|_2^2 + \alpha^T D(X^TX)^{-1}(X^Ty - D^T\alpha)$$

$$\min_\gamma\{\lambda\|\gamma\|_1 - \alpha^T\gamma\} = \begin{cases} 0, & \|\alpha\|_\infty \le \lambda \\ -\infty, & \text{otherwise} \end{cases}.$$

From $-X^T(y - X\beta) + D^T\alpha = 0$, we have $\beta = (X^TX)^{-1}(X^Ty - D^T\alpha)$. Thus, the minimum value can be written as

$$\frac{1}{2}\|(I - X(X^TX)^{-1}X^T)y + X(X^TX)^{-1}D^T\alpha)\|_2^2 + \alpha^T D(X^TX)^{-1}(X^Ty - D^T\alpha)$$

$$\sim \frac{1}{2}\{X(X^TX)^{-1}D^T\alpha\}^T X(X^TX)^{-1}D^T\alpha + \alpha^T D(X^TX)^{-1}X^Ty - \alpha^T D(X^TX)^{-1}D^T\alpha$$

$$\sim -\frac{1}{2}\|X(X^TX)^{-1}X^Ty - X(X^TX)^{-1}D^T\alpha\|_2^2,$$

where $A \sim B$ is the equivalent relation $A - B$, which is a constant irrespective of α. The dual problem amounts to the maximization of this value and the minimization of

$$\frac{1}{2}\|X(X^TX)^{-1}X^Ty - X(X^TX)^{-1}D^T\alpha\|_2^2$$

under $\|\alpha\|_\infty \le \lambda$. □

Exercises 47–61

47. Execute the following fused Lasso procedures.

(a) We wish to obtain the smooth output $\theta = (\theta_1, \ldots, \theta_N)$ that minimizes

$$\frac{1}{2}\sum_{i=1}^N(y_i - \theta_i)^2 + \lambda\sum_{i=2}^N|\theta_i - \theta_{i-1}| \quad \text{(cf. (4.1))} \quad (4.30)$$

from the observations $y = (y_1, \ldots, y_N)$, where $\lambda \ge 0$ is a parameter that controls the size of the output. The following procedure smooths the data of the copy number ratio of a gene in the chromosomal region between case and control patients that are measured by CGH (comparative genomic hybridization). Download the data from

https://web.stanford.edu/~hastie/StatLearnSparsity_files/DATA/cgh.html

fill in the blank, and execute the procedure to observe what smoothing procedure can be obtained for $\lambda = 0.1, 1, 10, 100, 1000$.

```
1  library(genlasso)
2  df = read.table("cgh.txt"); y = df[[1]]; N = length(y)
3  theta = ## Blank ##
4  plot(1:N,   theta, lambda = 0.1, xlab = "gene #", ylab = "Copy Number
        Variant",
5         col = "red", type = "l")
6  points(1:N, y, col = "blue")
```

(b) The fused Lasso can consider the differences w.r.t. the second-order $\theta_i - 2\theta_{i+1} + \theta_{i+2}$, w.r.t. the third-order $\theta_i - 3\theta_{i+1} + 3\theta_{i+2} - \theta_{i+3}$, etc., as well as w.r.t. the first order $\theta_i - \theta_{i+1}$ (trend filtering). Using the function trendfilter prepared in the genlasso package, we wish to execute trend filtering for the data $\sin\theta$ ($0 \le \theta \le 2\pi$) with random noise added. Determine the value of λ appropriately, and then execute trendfilter of the $k = 1, 2, 3$, 4-th order to show the graph similarly to (a).

```
1  library(genlasso)
2  ## Data Generation
3  N = 100; y = sin(1:N/(N * 2 * pi)) + rnorm(N, sd = 0.3)
4  out = ## Blank ##
5  plot(out, lambda = lambda)   ## Smoothing and Output
```

48. We wish to obtain the $\theta_1, \ldots, \theta_N$ that minimize (4.30) via dynamic programming from the observations $y_1, \ldots, y_N$.

(a) The condition of minimization w.r.t. θ_1

$$h_1(\theta_1, \theta_2) := \frac{1}{2}(y_1 - \theta_1)^2 + \lambda|\theta_2 - \theta_1|$$

contains another variable θ_2. The optimum θ_1 when the θ_2 value is known can be written as

$$\hat{\theta}_1(\theta_2) = \begin{cases} y_1 - \lambda, & y_1 \ge \theta_2 + \lambda \\ \theta_2, & |y_1 - \theta_2| < \lambda \\ y_1 + \lambda, & y_1 \le \theta_2 - \lambda \end{cases}.$$

The condition of minimization w.r.t. θ_2

$$\frac{1}{2}(y_1 - \theta_1)^2 + \frac{1}{2}(y_2 - \theta_2)^2 + \lambda|\theta_2 - \theta_1| + \lambda|\theta_3 - \theta_2|$$

contains other variables θ_1, θ_3. However, if we replace θ_1 with $\hat{\theta}_1(\theta_2)$, it can be written as

$$h_2(\hat{\theta}_1(\theta_2), \theta_2, \theta_3) := \frac{1}{2}(y_1 - \hat{\theta}_1(\theta_2))^2 + \frac{1}{2}(y_2 - \theta_2)^2 + \lambda|\theta_2 - \hat{\theta}_1(\theta_2)| + \lambda|\theta_3 - \theta_2|,$$

which is the $\hat{\theta}_2(\theta_3)$ value when the θ_3 value can be written as a function of θ_3. Show that the problem of finding $\theta_1, \ldots, \theta_N$ reduces to finding the θ_N value that minimizes

$$\frac{1}{2}\sum_{i=1}^{N-1}(y_i - \hat{\theta}_i(\theta_N))^2 + \frac{1}{2}(y_N - \theta_N)^2 + \lambda\sum_{i=1}^{N-2}|\hat{\theta}_i(\theta_N) - \hat{\theta}_{i+1}(\theta_N)| + \lambda|\hat{\theta}_{N-1}(\theta_N) - \theta_N|$$

$$\text{(cf. (4.4)) .}$$

(b) When we obtain θ_N in (a), how can we obtain the $\theta_1, \ldots, \theta_{N-1}$ values?

49. When we solve the fused Lasso via dynamic programming, the computation time is proportional to the length N of $y \in \mathbb{R}^N$. Explain the fundamental reason for this fact.

50. When we find the value of θ_i that minimizes

$$\frac{1}{2}\sum_{i=1}^{N}(y_i - \theta_i)^2 + \lambda\sum_{i=1}^{N}|\theta_i| + \mu\sum_{i=2}^{N}|\theta_i - \theta_{i-1}| \quad \text{(cf. (4.2)) ,} \qquad (4.31)$$

without loss of generality, it is sufficient to obtain only $\{\theta_i\}$ under $\lambda = 0$. In fact, for the solution $\theta(0)$ when $\lambda = 0$, $\mathcal{S}_\lambda(\theta(0))$ is the solution for general λ. In the following, for $y_1, y_2, \ldots, y_{N-1}, y_N \in \mathbb{R}$, $\lambda, \mu \geq 0$, we show the existence of

$$s_i(\lambda) = \begin{cases} 1, & \theta > 0 \\ [-1, 1], & \theta = 0 \\ -1, & \theta < 0 \end{cases}, \quad t_i(\lambda) = \begin{cases} 1, & \theta_i > \theta_{i-1} \\ [-1, 1], & \theta_i = \theta_{i-1} \\ -1, & \theta_i < \theta_{i+1} \end{cases}, \quad u_i(\lambda) = \begin{cases} 1, & \theta_i > \theta_{i+1} \\ [-1, 1], & \theta_i = \theta_{i+1} \\ -1, & \theta_i < \theta_{i+1} \end{cases}$$

such that

$$f_i(\lambda) := -y_i + \theta_i(\lambda) + \lambda s_i(\lambda) + \mu t_i(\lambda) + \mu u_i(\lambda) = 0 \quad (i = 2, \ldots, N-1)$$
$$\text{(cf. (4.25)),}$$

$$\theta_i(\lambda) = \mathcal{S}_\lambda(\theta_i(0)) \qquad \text{(cf. (4.26))}$$

where we assume that $(s_i(0), t_i(0), u_i(0))$ exists such that $f_i(0) = 0$.

(a) Show that $-y_i + \theta_i(0) + \mu t_i(0) + \mu u_i(0) = 0$.
(b) For $i, j = 1, \ldots, N$ $(i \neq j)$, show that $\theta_i(0) = \theta_j(0) \Longrightarrow \theta_i(\lambda) = \theta_j(\lambda)$ and $\theta_i(0) < \theta_j(0) \Longrightarrow \theta_i(\lambda) \leq \theta_j(\lambda)$.
(c) If we regard the solution $t_i(0), t_i(\lambda)$ as the set $(\{1\}, [-1, 1], \{-1\})$, show that $t_i(0) \subseteq t_i(\lambda)$.

(d) When $|\theta_i(0)| \geq \lambda$, show that $(s_i(\lambda), t_i(\lambda), u_i(\lambda)) = (s_i(0), t_i(0), u_i(0))$ is the solution of $f_i(\lambda) = 0$.

(e) When $|\theta_i(0)| < \lambda$, show that $(s_i(\lambda), t_i(\lambda), u_i(\lambda)) = (\theta_i(0)/\lambda, t_i(0), u_i(0))$ is the solution of $f_i(\lambda) = 0$.

(f) Let $\hat{\theta}_i(\lambda_1, \lambda_2)$ $(i = 1, \ldots, N)$ be the θ_i that minimizes (4.31). Show that $\hat{\theta}_i(\lambda_1, \lambda_2) = \mathcal{S}_{\lambda_1}(\hat{\theta}_i(0, \lambda_2))$.

51. When we minimize (4.31) based on Exercise 50, what process is required to obtain the solutions for $\lambda \neq 0$ after obtaining the solution for $\lambda = 0$? Specify the R language function.

52. In linear regression Lasso, there exists a smallest λ such that no variables are active. Let λ_0 be such a λ. In LARS, it is the maximum value of $\langle r_0, x_j \rangle$ when $r_0 := y$. Let S be the set of indices of the active variables. Initially, we have $S = j$. In LARS, if we decrease the λ value, only the coefficients of the active variables increase. Show that if an index j becomes active at λ,

$$< x^{(j)}, r(\lambda) >= \pm\lambda$$

for $\lambda' \leq \lambda$, where $r(\lambda)$ is the residual for λ and $x^{(j)}$ expresses the j-th column of the matrix X.

53. We construct the LARS function as follows. Fill in the blanks, and execute the procedure.

```
1   lars = function(X, y) {
2     X = as.matrix(X); n = nrow(X); p = ncol(X); X.bar = array(dim = p)
3     for (j in 1:p) {X.bar[j] = mean(X[, j]); X[, j] = X[, j] - X.bar[j]}
4     y.bar = mean(y); y = y - y.bar
5     scale = array(dim = p)
6     for (j in 1:p) {scale[j] = sqrt(sum(X[, j] ^ 2) / n); X[, j] = X[, j]
          / scale[j]}
7     beta = matrix(0, p + 1, p); lambda = rep(0, p + 1)
8     for (i in 1:p) {
9       lam = abs(sum(X[, i] * y))
10      if (lam > lambda[1]) {i.max = i; lambda[1] = lam}
11    }
12    r = y; index = i.max; Delta = rep(0, p)
13    for (k in 2:p) {
14      Delta[index] = solve(t(X[, index]) %*% X[, index]) %*%
15        t(X[, index]) %*% r / lambda[k - 1]
16      u = t(X[, -index]) %*% (r - lambda[k - 1] * X %*% Delta)
17      v = -t(X[, -index]) %*% (X %*% Delta)
18      t = ## Blank(1) ##
19      for (i in 1:(p - k + 1)) if (t[i] > lambda[k]) {lambda[k] = t[i]; i.
          max = i}
20      t = u / (v - 1)
21      for (i in 1:(p - k + 1)) if (t[i] > lambda[k]) {lambda[k] = t[i]; i.
          max = i}
22      j = setdiff(1:p, index)[i.max]
23      index = c(index, j)
24      beta[k, ] = ## Blank(2) ##
25      r = y - X %*% beta[k, ]
26    }
27    for (k in 1:(p + 1)) for (j in 1:p) beta[k, j] = beta[k, j] / scale[j]
```

```
28        return(list(beta = beta, lambda = lambda))
29    }
30    df = read.table("crime.txt"); X = as.matrix(df[, 3:7]); y = df[, 1]
31    res = lars(X, y)
32    beta = res$beta; lambda = res$lambda
33    p = ncol(beta)
34    plot(0:8000, ylim = c(-7.5, 15), type = "n",
35         xlab = "lambda", ylab = "beta", main = "LARS(USA Crime Data)")
36    abline(h = 0) for (j in 1:p)
37    lines(lambda[1:(p)], beta[1:(p), j], col = j)
38    legend("topright",
39             legend = c("annual police funding", "people 25 years+ with 4 yrs.
             of high school",
40                         "16--19 year-olds not in highschool and not highschool
             graduates",
41                         "18--24 year-olds in college",
42                         "people 25 years+ with at least 4 years of college"),
43           col = 1:p, lwd = 2, cex = .8)
```

54. From fused Lasso, we consider the extended formulation (generalized Lasso): for $X \in \mathbb{R}^{N \times p}$, $y \in \mathbb{R}^N$, $\beta \in \mathbb{R}^p$, and $D \in \mathbb{R}^{m \times p}$ ($m \leq p$), we minimize

$$\frac{1}{2}\|y - X\beta\|^2 + \lambda\|D\beta\|_1 . \tag{4.32}$$

Why is the ordinary linear regression Lasso a particular case of (4.32)? How about the ordinary fused Lasso? What D should be given for the two cases below (trend filtering)?

i. $\dfrac{1}{2}\displaystyle\sum_{i=1}^{N}(y_i - \theta_i)^2 + \displaystyle\sum_{i=1}^{N-2}|\theta_i - 2\theta_{i+1} + \theta_{i+2}|$

ii. $\dfrac{1}{2}\displaystyle\sum_{i=1}^{N}(y_i - \theta_i)^2 + \displaystyle\sum_{i=1}^{N-2}\left|\dfrac{\theta_{i+2} - \theta_{i+1}}{x_{i+2} - x_{i+1}} - \dfrac{\theta_{i+1} - \theta_i}{x_{i+1} - x_i}\right|$ (cf. (4.13))   (4.33)

55. Derive the dual problem from each of the primary problems below.

(a) For $X \in \mathbb{R}^{N \times p}$, $y \in \mathbb{R}^N$, and $\lambda > 0$, find the $\beta \in \mathbb{R}^p$ that minimizes

$$\frac{1}{2}\|y - X\beta\|_2^2 + \lambda\|\beta\|_1 \qquad \text{(cf. (4.7))}$$

(Primary). Under $\|X^T\alpha\|_\infty \leq \lambda$, find the $\alpha \in \mathbb{R}^N$ that maximizes

$$\frac{1}{2}\{\|y\|_2^2 - \|y - \alpha\|_2^2\}$$

(Dual).

(b) For $\lambda \geq 0$ and $D \in \mathbb{R}^{m \times N}$, find the $\theta \in \mathbb{R}^N$ that minimizes

$$\frac{1}{2}\|y - \theta\|_2^2 + \lambda\|D\theta\|_1$$

(Primary). Under $\|\alpha\|_\infty \leq \lambda$, find the $\alpha \in \mathbb{R}^N$ that minimizes

$$\frac{1}{2}\|y - D^T\alpha\|_2^2$$

(Dual).

(c) For $X \in \mathbb{R}^{N \times p}$, $y \in \mathbb{R}^N$, and $D \in \mathbb{R}^{m \times p}$ ($m \geq 1$), find the $\beta \in \mathbb{R}^p$ that minimizes

$$\frac{1}{2}\|y - X\beta\|_2^2 + \lambda\|D\beta\|_1$$

(Primary). Under $\|\alpha\|_\infty \leq \lambda$, find the $\alpha \in \mathbb{R}^m$ that minimizes

$$\frac{1}{2}\|X(X^T X)^{-1}y - X(X^T X)^{-1}D^T\alpha\|_2^2 \qquad \text{(cf. (4.15))}$$

(Dual).

56. Suppose that we solve fused Lasso as a dual problem. Let λ_1 and i_1 be the largest absolute value and its element i among the elements of the solution $\hat{\alpha}^{(1)}$ of $\|y - D^T\alpha\|_2^2$. For $k = 2, \ldots, m$, we execute the following procedure to define the sequences $\{\lambda_k\}, \{i_k\}, \{s_k\}$. For $\mathcal{S} := \{i_1, \ldots, i_{k-1}\}$, let λ_k and i_k be the largest absolute value and its element i among the elements of the solution $\alpha_{-\mathcal{S}}$ that minimizes

$$\frac{1}{2}\|y - \lambda D_{\mathcal{S}}^T s - D_{-\mathcal{S}}^T\alpha_{-\mathcal{S}}\|_2^2 \qquad \text{(cf. (4.19))} .$$

If $\hat{\alpha}_{i_k} = \lambda_k$, then $s_{i_k} = 1$ and $s_{i_k} = -1$ otherwise, where $D_{\mathcal{S}} \in \mathbb{R}^{k \times p}$, $\alpha_{\mathcal{S}} \in \mathbb{R}^k$ consists of the rows that correspond to $\mathcal{S}$ and $D_{-\mathcal{S}} \in \mathbb{R}^{(m-k) \times p}$, $\alpha_{-\mathcal{S}} \in \mathbb{R}^{m-k}$ consists of the other rows. If we differentiate it by $\alpha_{-\mathcal{S}}$ to minimize it, the solution is

$$\alpha_k(\lambda) := \{D_{-\mathcal{S}}D_{-\mathcal{S}}^T\}^{-1}D_{-\mathcal{S}}(y - \lambda D_{\mathcal{S}}^T s) \qquad \text{(cf. (4.20))}\cdot$$

If we set the $i \notin \mathcal{S}$-th element as $a_i - \lambda b_i$, what are a_i, b_i? Moreover, how can we obtain i_k, λ_k?

57. We construct a program that obtains the solutions of the dual and primary fused Lasso problems. Fill in the blanks, and execute the procedure.

```
1   fused.dual = function(y, D) {
2     m = nrow(D)
3     lambda = rep(0, m); s = rep(0, m); alpha = matrix(0, m, m)
4     alpha[1, ] = solve(D %*% t(D)) %*% D %*% y
5     for (j in 1:m) if (abs(alpha[1, j]) > lambda[1]) {
6       lambda[1] = abs(alpha[1, j])
7       index = j
8       if (alpha[1, j] > 0) ## Blank(1) ##
9     }
```

```
10    for (k in 2:m) {
11      U = solve(D[-index, ] %*% t(as.matrix(D[-index, , drop = FALSE])))
12      V = D[-index, ] %*% t(as.matrix(D[index, , drop = FALSE]))
13      u = U %*% D[-index, ] %*% y
14      v = U %*% V %*% s[index]
15      t = u / (v + 1)
16      for (j in 1:(m - k + 1)) if (t[j] > lambda[k]) {lambda[k] = t[j]; h
        = j; r = 1}
17      t = u / (v - 1)
18      for (j in 1:(m - k + 1)) if (t[j] > lambda[k]) {lambda[k] = t[j]; h
        = j; r = -1}
19      alpha[k, index] = ## Blank(2) ##
20      alpha[k, -index] = ## Blank(3) ##
21      h = setdiff(1:m, index)[h]
22      if (r == 1) s[h] = 1 else s[h] = -1
23      index = c(index, h)
24    }
25    return(list(alpha = alpha, lambda = lambda))
26  }
27  m = p - 1; D = matrix(0, m, p); for (i in 1:m) {D[i, i] = 1; D[i, i + 1]
      = -1}
28  fused.prime = function(y, D){
29    res = fused.dual(y, D)
30    return(list(beta = t(y - t(D) %*% t(res$alpha)), lambda = res$lambda))
31  }
32  p = 8; y = sort(rnorm(p)); m = p - 1; s = 2 * rbinom(m, 1, 0.5) - 1
33  D = matrix(0, m, p); for (i in 1:m) {D[i, i] = s[i]; D[i, i + 1] = -s[i
      ]}
34  par(mfrow = c(1, 2)) res = fused.dual(y, D); alpha =
35  res$alpha; lambda = res$lambda
36  lambda.max = max(lambda); m =
37  nrow(alpha) alpha.min = min(alpha); alpha.max = max(alpha)
38  plot(0:lambda.max, xlim = c(0, lambda.max), ylim = c(alpha.min, alpha.
        max),
39        type = "n", xlab = "lambda", ylab = "alpha", main = "Dual Problem")
40  u = c(0, lambda); v = rbind(0, alpha); for (j in 1:m) lines(u, v[,
41  j], col = j); res = fused.prime(y, D); beta = res$beta
42  beta.min = min(beta); beta.max = max(beta)
43  plot(0:lambda.max, xlim = c(0, lambda.max), ylim = c(beta.min, beta.max
        ),
44        type = "n", xlab = "lambda", ylab = "beta", main = "Primary Problem
        ")
45  w = rbind(0, beta); for (j in 1:p) lines(u, w[, j], col = j)
46  par(mfrow = c(1, 1))
```

58. If the design matrix is not the unit matrix for the generalized Lasso, then for $X^+ := (X^T X)^{-1} X^T$, $\tilde{y} := X X^+ y$, $\tilde{D} := D X^+$, the problem reduces to minimizing

$$\frac{1}{2}\|\tilde{y} - \tilde{D}\alpha\|_2^2 .$$

Extend the functions `fused.dual` and `fused.prime` to `fused.dual.general` and `fused.prime.general`, and execute the following procedure.

```
1   n = 20; p = 10; beta = rnorm(p + 1)
2   X = matrix(rnorm(n * p), n, p); y = cbind(1, X) %*% beta + rnorm(n)
3   # D = diag(p)   ## Use either
4   D = array(dim = c(p - 1, p))
5   for (i in 1:(p - 1)) {D[i, ] = 0; D[i, i] = 1; D[i, i + 1] = -1}
6   par(mfrow = c(1, 2))
7   res = fused.dual.general(X, y, D); alpha = res$alpha; lambda = res$
        lambda
8   lambda.max = max(lambda); m = nrow(alpha)
9   alpha.min = min(alpha); alpha.max = max(alpha)
10  plot(0:lambda.max, xlim = c(0, lambda.max), ylim = c(alpha.min, alpha.
        max),
11      type = "n", xlab = "lambda", ylab = "alpha", main = "Dual Problem")
12  u = c(0, lambda); v = rbind(0, alpha); for (j in 1:m) lines(u, v[, j],
        col = j)
13  res = fused.prime.general(X, y, D); beta = res$beta
14  beta.min = min(beta); beta.max = max(beta)
15  plot(0:lambda.max, xlim = c(0, lambda.max), ylim = c(beta.min, beta.max
        ),
16      type = "n", xlab = "lambda", ylab = "beta", main = "Primary Problem
        ")
17  w = rbind(0, beta); for (j in 1:p) lines(u, w[, j], col = j)
18  par(mfrow = c(1, 1))
```

59. Set $A \in \mathbb{R}^{d \times m}$, $B \in \mathbb{R}^{d \times n}$, and $c \in \mathbb{R}^d$. We assume that $f : \mathbb{R}^m \to \mathbb{R}$ and $g : \mathbb{R}^n \to \mathbb{R}$ are convex and that f is differentiable. We formulate the problem that minimizes $f(\alpha) + g(\beta)$ under $A\alpha + B\beta = c$ by adding a constant $\rho > 0$ to the Lagrange multiplier

$$f(\alpha) + g(\beta) + \gamma^T(A\alpha + B\beta - c) \to \min \quad (\gamma \in \mathbb{R}^d)$$

to write

$$L_\rho(\alpha, \beta, \gamma) := f(\alpha) + g(\beta) + \gamma^T(A\alpha + B\beta - c) + \frac{\rho}{2}\|A\alpha + B\beta - c\|^2$$

$$\text{(cf. (4.23))} .$$

If we repeat the updates expressed by the three equations, we can obtain the optimum solution under some conditions (ADMM, alternating direction method of multipliers).

$$\begin{cases} \alpha_{t+1} \leftarrow \alpha \in \mathbb{R}^m \text{ that minimizes } L_\rho(\alpha, \beta_t, \gamma_t) \\ \beta_{t+1} \leftarrow \beta \in \mathbb{R}^n \text{ that minimizes } L_\rho(\alpha_{t+1}, \beta, \gamma_t) \\ \gamma_{t+1} \leftarrow \gamma_t + \rho(A\alpha_{t+1} + B\beta_{t+1} - c) \end{cases}$$

Suppose that we apply similar updates to the generalized Lasso:

$$L_\rho(\theta, \gamma, \mu) := \frac{1}{2}\|y - \theta\|_2^2 + \lambda\|\gamma\|_1 + \mu^T(D\theta - \gamma) + \frac{\rho}{2}\|D\theta - \gamma\|^2$$

Answer the following questions.

(a) What are $A \in \mathbb{R}^{d \times m}$, $B \in \mathbb{R}^{d \times n}$, $c \in \mathbb{R}^d$, $f : \mathbb{R}^m \to \mathbb{R}$, and $g : \mathbb{R}^n \to \mathbb{R}$?

(b) Show that the updates are given as follows:

$$\begin{cases} \theta_{t+1} \leftarrow (I + \rho D^T D)^{-1}(y + D^T(\rho\gamma_t - \mu_t)) \\ \gamma_{t+1} \leftarrow S_\lambda(\rho D\theta_{t+1} + \mu_t)/\rho \\ \mu_{t+1} \leftarrow \mu_t + \rho(D\theta_{t+1} - \gamma_{t+1}) \end{cases}$$

Hint

$$\frac{\partial L_\rho}{\partial \theta} = \theta - y + D^T\mu_t + \rho D^T(D\theta - \gamma_t)$$

$$\frac{\partial L_\rho}{\partial \gamma} = -\mu_t + \rho(\gamma - D\theta_{t+1}) + \lambda \begin{cases} 1, & \gamma > 0 \\ [-1, 1], & \gamma = 0 \\ -1, & \gamma < 0 \end{cases}$$

60. Fill in the blanks below, and construct the function admm that realizes the generalized Lasso.

```
1   admm = function(y, D, lambda) {
2     K = ncol(D); L = nrow(D)
3     theta.old = rnorm(K); theta = rnorm(K); gamma = rnorm(L); mu = rnorm(L
        )
4     rho = 1
5     while (max(abs(theta - theta.old) / theta.old) > 0.001) {
6       theta.old = theta
7       theta = ## Blank(1) ##
8       gamma = ## Blank(2) ##
9       mu = mu + ## Blank(3) ##
10    }
11    return(theta)
12  }
```

61. Using Exercise 60, for each of the following cases, fill in the blanks to execute the procedure.

(a) Lasso in (4.33) (trend filtering). The data can be downloaded from

https://web.stanford.edu/~hastie/StatLearnSparsity_files/DATA/
airPollution.txt

```
1    ## Vector Input df = read.table("airpolution.txt", header = TRUE)
2    index = order(df[[3]])
3    y = df[[1]][index]; N = length(y)
4    x = df[[3]] + rnorm(N) * 0.01
5    # The original data contains the same integer values and some
         purterbation was added
6    x = x[index] ## Setting Matrix D
7    D = matrix(0, ncol = N, nrow = N - 2)
8    for (i in 1:(N - 2)) D[i, ] = 0
9    for (i in 1:(N - 2)) D[i, i] = 1 / (x[i + 1] - x[i])
10   for (i in 1:(N - 2)) D[i, i + 1] = -1 / (x[i + 1] - x[i]) - 1 / (x[i
         + 2] - x[i + 1])
11   for (i in 1:(N - 2)) D[i, i + 2] = ## Blank(1) ##
12   ## Computation of theta
```

```
13  theta = ## Blank(2) ##
14  plot(x, theta, xlab = "Temperature (F)", ylab = "Ozon", col = "red",
           type = "l")
15  points(x, y, col = "blue")
```

(b) See the solution path merge in the fused Lasso, where we use the CGH data
     in Exercise 47 (a).

```
1   df = read.table("cgh.txt"); y = df[[1]][101:110]; N = length(y)
2   D = array(dim = c(N - 1, N)) for (i in 1:(N - 1)) {D[i, ] = 0; D[i,
        i] =
3   1; D[i, i + 1] = -1}
4   lambda.seq = seq(0, 0.5, 0.01); M = length(lambda.seq)
5   theta = list(); for (k in 1:M) theta[[k]] = ## Blank(3) ##
6   x.min = min(lambda.seq); x.max = max(lambda.seq)
7   y.min = min(theta[[1]]); y.max = max(theta[[1]])
8   plot(lambda.seq, xlim = c(x.min, x.max), ylim = c(y.min, y.max),
           type = "n",
9           xlab = "lambda", ylab = "Coefficient", main = "Fused Lasso
           Solution Path")
10  for (k in 1:N) {
11    value = NULL; for (j in 1:M) value = c(value, theta[[j]][k])
12    lines(lambda.seq, value, col = k)
13  }
```

Chapter 5
Graphical Models

In this chapter, we examine the problem of estimating the structure of the graphical model from observations. In the graphical model, each vertex is regarded as a variable, and edges express the dependency between them (conditional independence). In particular, assume a so-called sparse situation where the number of vertices is larger than the number of variables . Consider the problem of connecting vertex pairs with a certain degree of dependency or more as edges. In the structural estimation of a graphical model assuming sparsity, the so-called undirected graph with no edge orientation is often used.

In this chapter, we first introduce the theory of graphical models, especially the concept of conditional independence and the separation of undirected graphs. Then, we learn the algorithms of graphical Lasso, structural estimation of the graphical model using the pseudo-likelihood, and joint graphical Lasso.

5.1 Graphical Models

The graphical model dealt with in this book is related to undirected graphs. For $p \geq 1$, we define $V := \{1, \ldots, p\}$, and E is a subset of $\{\{i, j\} \mid i \neq j, \ i, j \in V\}$.[1] In particular, V is called a vertex set, its elements are called vertices, E is called an edge set, and its elements are called edges. The undirected graph consisting of them is written as (V, E). For subsets A, B, C of V, when all paths connecting A, B pass through some vertex of C, C is said to separate A and B and is written as $A \perp\!\!\!\perp_E B \mid C$ (Fig. 5.1).

---

[1] We do not distinguish between $\{1, 2\}$ and $\{2, 1\}$.

© The Author(s), under exclusive license to Springer Nature Singapore Pte Ltd. 2021
J. Suzuki, *Sparse Estimation with Math and R*,
https://doi.org/10.1007/978-981-16-1446-0_5

The red verteces separate the blue and green verteces.

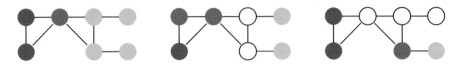

The red verteces do not separate the blue and green verteces.

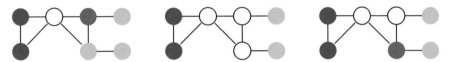

Fig. 5.1 Vertex sets A, B are separated by another vertex set. The blue and green vertices are separated and not separated by red ones above and below the three undirected graphs, respectively. There is a red vertex on any path connecting green and blue vertices in each of the three above graphs, while there is at least one path connecting green and blue vertices that do not contain any red vertex in each of the three below graphs. For example, in the left-below graph, the two left blue vertices can reach the green vertex without passing through the red vertex. The two blue and two green vertices communicate with each other without passing through the red vertex in the center-below graph. In the right-below graph, the two blue and one green vertices communicate without passing through the red vertex

In the following, we identify p random variables $X_1, \ldots, X_p$ and vertices $1, \ldots, p$ and express their conditional independence properties by the undirected graph (V, E).

For variables X, Y, we write the probabilities of the events associated with $X, Y, (X, Y)$ as $P(X)$, $P(Y)$, and $P(X, Y)$, respectively. We say that X, Y are independent and write $X \perp\!\!\!\perp_P Y$ if $P(X)P(Y) = P(X, Y)$. Similarly, we write the probability of the event associated with (X, Y, Z) as $P(X, Y, Z)$. We say that X, Y are conditionally independent given Z and write $X \perp\!\!\!\perp_P Y \mid Z$ if $P(X, Z)P(Y, Z) = P(X, Y, Z)P(Z)$.

A graphical model expresses multiple conditional independence (CI) relations by the following graph: for disjoint subsets X_A, X_B, X_C of $\{X_1, \ldots, X_p\}$, the edges are connected such that

$$X_A \perp\!\!\!\perp_P X_B \mid X_C \iff A \perp\!\!\!\perp_E B \mid C . \tag{5.1}$$

However, in general, such an edge set E of (V, E) may not exist for some probabilities P.

Example 45 Let X, Y be random variables that take zeros and ones equiprobably, and assume that Z is the residue of $X + Y$ when divided by two. Because (X, Y) takes four values equiprobably, we have

$$P(X)P(Y) = P(X, Y)$$

and find that X, Y are independent. However, each of $(X, Z), (Y, Z), (X, Y, Z)$ takes four values equiprobably, and Z takes two values equiprobably. Thus, we have

$$P(X, Z)P(Y, Z) = \frac{1}{4} \cdot \frac{1}{4} \neq \frac{1}{4} \cdot \frac{1}{2} = P(X, Y, Z)P(Z),$$

which means that X, Y are not conditionally independent given Z. From the implication $\Longleftarrow$ in (5.1), we find that X, Z, Y are not connected in this order. If we know the value of Z, then we also know whether $X = Y$ or $X \neq Y$, which means that they are not independent given Z. In addition, both $(X, Z), (Y, Z)$ are not independent, and they should be connected as edges. Hence, we need to connect all three edges, although the graph cannot express $X \perp\!\!\!\perp_P Y$. Therefore, we cannot distinguish by any undirected graph between the CI relation in this example and the case where no nontrivial CI relation exists.

However, if we connect all the edges, the implication $\Longleftarrow$ in (5.1) trivially holds. We say that an undirected graph for probability P is a Markov network of P if the number of edges is minimal and the implication $\Longleftarrow$ in (5.1) holds.

However, for the Gaussian variables, we can obtain the CI properties from the covariance matrix $\Sigma \in \mathbb{R}^{p \times p}$. For simplicity, we assume that the means of the variables are zeros. Let Θ be the inverse matrix of Σ, and write the probability density matrix as

$$f(x) = \sqrt{\frac{\det \Theta}{(2\pi)^p}} \exp\left\{-\frac{1}{2}x^T \Theta x\right\} \quad (x \in \mathbb{R}^p).$$

Then, the sufficient and necessary condition for disjoint subsets A, B, C of $\{1, \ldots, p\}$ to be conditionally independent is

$$f_{A \cup C}(x_{A \cup C}) f_{B \cup C}(x_{B \cup C}) = f_{A \cup B \cup C}(x_{A \cup B \cup C}) f_C(x_C) \tag{5.2}$$

for arbitrary $x \in \mathbb{R}^p$, which is equivalent to

$$
\begin{aligned}
&- \log \det \Sigma_{A \cup C} - \log \det \Sigma_{B \cup C} + \log \det \Sigma_{A \cup B \cup C} + \log \det \Sigma_C \\
&= x_{A \cup C}^T \Theta_{A \cup C} x_{A \cup C} + x_{B \cup C}^T \Theta_{B \cup C} x_{B \cup C} - x_{A \cup B \cup C}^T \Theta_{A \cup B \cup C} x_{A \cup B \cup C} - x_C^T \Theta_C x_C
\end{aligned}
\tag{5.3}
$$

for arbitrary $x \in \mathbb{R}^p$, where Θ_S, x_S are elements of Θ, x that correspond to a set $S \subseteq \{1, \ldots, p\}$, respectively. Among $x_{A \cup B \cup C}^T \Theta_{A \cup B \cup C} x_{A \cup B \cup C}$ on the right-hand side, the values of $x_i \Theta_{i,j} x_j$ ($i \in A$, $j \in B$) should be zeros because they are missing on the left-hand side. Thus, we require the elements Θ_{AB}, Θ_{BA} of Θ indexed by $i \in A$, $j \in B$ and $i \in B$, $j \in A$ to be zero. On the other hand, if $\Theta_{AB} = \Theta_{BA} = 0$, then both sides of (5.3) are zero (Proposition 15).

Proposition 15 *(Lauritzen, 1996 [18])*

$$\Theta_{AB} = \Theta_{BA}^T = 0 \implies \det \Sigma_{A \cup C} \det \Sigma_{B \cup C} = \det \Sigma_{A \cup B \cup C} \det \Sigma_C$$

For the proof, see the Appendix.

Example 46 Because $\Theta = \begin{bmatrix} \theta_{1,1} & 0 & \theta_{1,3} \\ 0 & \theta_{2,2} & \theta_{2,3} \\ \theta_{1,3} & \theta_{2,3} & \theta_{3,3} \end{bmatrix}$ implies

$$\Sigma = \frac{1}{\det \Theta} \begin{bmatrix} \theta_{2,2}\theta_{3,3} - \theta_{2,3}^2 & \theta_{1,3}\theta_{2,3} & -\theta_{1,3}\theta_{2,2} \\ \theta_{1,3}\theta_{2,3} & \theta_{1,1}\theta_{3,3} - \theta_{1,3}^2 & -\theta_{2,3}\theta_{1,1} \\ -\theta_{1,3}\theta_{2,2} & -\theta_{2,3}\theta_{1,1} & \theta_{1,1}\theta_{2,2} \end{bmatrix},$$

we have

$\det \Sigma_{\{1,3\}} \det \Sigma_{\{2,3\}}$

$$= \frac{1}{(\det \Theta)^2} \{(\theta_{2,2}\theta_{3,3} - \theta_{2,3}^2)\theta_{1,1}\theta_{2,2} - \theta_{1,3}^2\theta_{2,2}^2\} \cdot \frac{1}{(\det \Theta)^2} \{(\theta_{1,1}\theta_{3,3} - \theta_{1,3}^2)\theta_{1,1}\theta_{2,2} - \theta_{2,3}^2\theta_{1,1}^2\}$$

$$= \frac{1}{(\det \Theta)^4} \theta_{1,1}\theta_{2,2}(\theta_{1,1}\theta_{2,2}\theta_{2,3} - \theta_{1,1}\theta_{2,3}^2 - \theta_{2,2}\theta_{1,3}^2)^2$$

$$= \frac{\theta_{1,1}\theta_{2,2}}{(\det \Theta)^2} = \det \Sigma_{\{3\}} \det \Sigma_{\{1,2,3\}}$$

Hence, if $X_1, \ldots, X_p$ are Gaussian, we have $\Theta_{AB} = \Theta_{BA}^T = 0 \iff$ (5.3). Thus, we have the following proposition.

Proposition 16 If $X_1, \ldots, X_p$ are Gaussian, we have

$$\Theta_{AB} = \Theta_{BA}^T = 0 \iff C \subseteq \{1, 2, \ldots, p\} \text{ exists such that } X_A \perp\!\!\!\perp_P X_B \mid X_C .$$

In particular, for $D \cap (A \cup B) = \phi$, we have

$$X_A \perp\!\!\!\perp_P X_B \mid X_C, \ C \subseteq D \implies X_A \perp\!\!\!\perp_P X_B \mid X_D .$$

As we have seen in Example 45, $X \perp\!\!\!\perp_P Y$ does not mean that $X \perp\!\!\!\perp_P Y \mid Z$. However, for the Gaussian distributions, from Proposition 16, no such example exists. The proposition is useful in this chapter (Fig. 5.2).

Proposition 17 *(Pearl and Paz, 1985)* Suppose that we construct an undirected graph (V, E) such that $\{i, j\} \in E \iff X_i \perp\!\!\!\perp_P X_j \mid X_S$ for $i, j \in V$ and $S \subseteq V$. Then, $X_A \perp\!\!\!\perp_P X_B \mid X_C \iff A \perp\!\!\!\perp_E B \mid C$ for all disjoint subsets A, B, C of V if and only if the probability P satisfies the following conditions:

$$X_A \perp\!\!\!\perp_P X_B \mid X_C \iff X_B \perp\!\!\!\perp_P X_A \mid X_C \tag{5.4}$$

Markov network

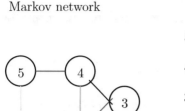

Precision Matrix Θ

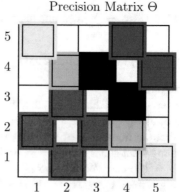

Fig. 5.2 In Proposition 17, for the Gaussian variables, the existence of each edge in the Markov network is equivalent to the nonzero of the corresponding element in the precision matrix. The colors red, blue, green, yellow, black, and pink in the left and right correspond

$$X_A \perp\!\!\!\perp_P (X_B \cup X_D) \mid X_C \implies X_A \perp\!\!\!\perp_P X_B | X_C, X_A \perp\!\!\!\perp_P X_D \mid X_C \tag{5.5}$$

$$X_A \perp\!\!\!\perp_P (X_C \cup X_D) \mid X_B, X_A \perp\!\!\!\perp_E X_D \mid (X_B \cup X_C) \implies X_A \perp\!\!\!\perp_E (X_B \cup X_D) \mid X_C \tag{5.6}$$

$$X_A \perp\!\!\!\perp_P X_B \mid X_C \implies X_A \perp\!\!\!\perp_P X_B \mid (X_C \cup X_D) \tag{5.7}$$

$$X_A \perp\!\!\!\perp_P X_B \mid X_C \implies X_A \perp\!\!\!\perp_P X_k \mid X_C \text{ or } X_k \perp\!\!\!\perp_E X_B | X_C \tag{5.8}$$

for $k \notin A \cup B \cup C$, where A, B, C, D are disjoint subsets of V for each equation.

For the proof, see the Appendix.

We claim that the Gaussian P satisfies (5.4)–(5.8), which means that if we construct an undirected graph (V, E) such that $\{i, j\} \in E \iff X_i \perp\!\!\!\perp_P X_j \mid X_S$ for $i, j \in V$ and $S \subseteq V$, then $X_A \perp\!\!\!\perp_P X_B \mid X_C \iff A \perp\!\!\!\perp_E B \mid C$.

Among the five relations, the first three hold for any probability model. The fourth is true for the Gaussian distributions from Proposition 16. We only show that the last relation holds for the Gaussian distribution. The relation if either $k \in A$ or $k \in B$, then the claim is apparent. If neither $X_A \perp\!\!\!\perp X_k | C$ nor $X_B \perp\!\!\!\perp X_k | C$, then it contradicts $X_A \perp\!\!\!\perp A_B | X_C$ because the distribution is Gaussian.

In this chapter, we consider estimating the edge set E of the Markov network from the observation $X \in \mathbb{R}^{N \times p}$ consisting of N tuples of the p variable $X_1, \ldots, X_p$. According to Proposition 17, we need to estimate only the precision matrix Θ, the inverse of the covariance matrix. If we write each element of the observation X as

$x_{i,j}$, then for $\bar{x}_j := \dfrac{1}{N} \displaystyle\sum_{i=1}^{N} x_{i,j}$, one might consider computing the sample covariance

matrix $S = (s_{i,j})$ whose elements are $s_{i,j} := \dfrac{1}{N} \displaystyle\sum_{i=1}^{N} (x_{i,j} - \bar{x}_j)^2$ and testing whether

each is zero or not.

However, we immediately find that the process causes problems. First, the matrix S does not always have the inverse. In particular, under the sparse situation, we assume that the number p of variables is large relative to the number N, such as when finding what genes among $p = 10,000$ affect the sick from p gene expression data of $N = 100$ cases/controls. S is nonnegative definite and can be expressed as $A^T A$ using a matrix $A \in \mathbb{R}^{N \times p}$. If $p > N$, then the rank of $S \in \mathbb{R}^{p \times p}$ is at most $N \ (< p)$, and S is not nonsingular.

Even if $p < N$, because we estimate Θ from samples, it is unlikely that all the elements of S_{AB} that correspond to Θ_{AB} happen to be zero, even if $X_A \perp\!\!\!\perp X_B \mid X_C$. Thus, we cannot infer the correct conclusion without applying anything similar to statistical testing.

5.2 Graphical Lasso

In the previous section, under $p > N$, we observe that no correct estimation of the CI relations among $X_1, \ldots, X_p$ is obtained even if we correctly estimate the sample covariance matrix S. In this section, using Lasso, we consider distinguishing whether each element $\Theta_{i,j}$ is zero or not (graphical Lasso [12]).

We first note that if we express the Gaussian density function of p variables as $f_\Theta(x_1, \ldots, x_p)$, the log-likelihood can be computed as

$$
\frac{1}{N} \sum_{i=1}^{N} \log f_\Theta(x_{i,1}, \ldots, x_{i,p})
$$

$$
= \frac{1}{2} \log \det \Theta - \frac{p}{2} \log(2\pi) - \frac{1}{2N} \sum_{i=1}^{N} \sum_{j=1}^{p} \sum_{k=1}^{p} x_{i,j} \theta_{j,k} x_{i,k}
$$

$$
= \frac{1}{2} \{ \log \det \Theta - p \log(2\pi) - \operatorname{trace}(S\Theta) \} , \tag{5.9}
$$

where we have used

$$
\frac{1}{N} \sum_{i=1}^{N} \sum_{j=1}^{p} \sum_{k=1}^{p} x_{i,j} \theta_{j,k} x_{i,k} = \sum_{j=1}^{p} \sum_{k=1}^{p} \theta_{j,k} \frac{1}{N} \sum_{i=1}^{N} x_{i,j} x_{i,k} = \sum_{j=1}^{p} \sum_{k=1}^{p} s_{j,k} \theta_{j,k} .
$$

Next, we L1-regularize (5.9) with $\lambda \geq 0$. Note that the maximization of

$$\frac{1}{N}\sum_{i=1}^{N}\log f_{\Theta}(x_i) - \frac{1}{2}\lambda\sum_{j\neq k}|\theta_{j,k}|$$

w.r.t. nonnegative definite Θ is equivalent to

$$\log\det\Theta - \text{trace}(S\Theta) - \lambda\sum_{j\neq k}|\theta_{j,k}| . \tag{5.10}$$

In the following, for simplicity, we write the determinant of a matrix A as either $|A|$ or det A.

In general, if we denote by $A_{i,j}$ the matrix that subtracts the i-th row and j-th column from matrix $A \in \mathbb{R}^{p\times p}$, then we have

$$\sum_{j=1}^{p}(-1)^{k+j}a_{i,j}|A_{k,j}| = \begin{cases} |A|, & i = k \\ 0, & i \neq k \end{cases} . \tag{5.11}$$

Suppose $|A| \neq 0$. If B is the matrix whose (j, k) element is $b_{j,k} = (-1)^{k+j}|A_{k,j}|/|A|$, then AB is the unit matrix. In fact, from (5.11), we have

$$\sum_{j=1}^{p}a_{i,j}b_{j,k} = \sum_{j=1}^{p}a_{i,j}(-1)^{k+j}\frac{|A_{k,j}|}{|A|} = \begin{cases} 1, & i = k \\ 0, & i \neq k \end{cases} .$$

Hereafter, we denote such a matrix B as A^{-1}. Then, we note that the differentiation of $|A|$ w.r.t. $a_{i,j}$ is $(-1)^{i+j}|A_{i,j}|$. In fact, we observe that if we let $i = k$ in (5.11), then the coefficient of $a_{i,j}$ in $|A|$ is $(-1)^{i+j}|A_{i,j}|$. In addition, if we differentiate $\text{trace}(S\Theta) = \sum_{i=1}^{p}\sum_{j=1}^{p}s_{i,j}\theta_{i,j}$ with the (k, h)-th element $\theta_{k,h}$ of Θ, it becomes the (k, h)-th element $s_{k,h}$ of S. Moreover, if we differentiate $\log\det\Theta$ by $\theta_{k,h}$, we have

$$\frac{\partial\log|\Theta|}{\partial\theta_{k,h}} = \frac{1}{|\Theta|}\frac{\partial|\Theta|}{\partial\theta_{k,h}} = \frac{1}{|\Theta|}(-1)^{k+h}|\Theta_{k,h}| ,$$

which is the (h, k)-th element of Θ^{-1}. Because Θ is symmetric, Θ^{-1} is also symmetric:

$$\Theta^T = \Theta \implies (\Theta^{-1})^T = (\Theta^T)^{-1} = \Theta^{-1} .$$

Thus, the Θ that maximizes (5.10) is the solution of

$$\Theta^{-1} - S - \lambda\Psi = 0 , \tag{5.12}$$

where $\Psi = (\psi_{j,k})$ is $\psi_{j,k} = 0$ when $j = k$ and

$$\psi_{j,k} = \begin{cases} 1, & \theta_{j,k} > 0 \\ [-1, 1], & \theta_{j,k} = 0 \\ -1, & \theta_{j,k} < 0 \end{cases}$$

otherwise.

The solution can be obtained from formulating the Lagrange multiplier of the nonnegative definite problem and applying the KKT condition:

$$-\log \det \Theta + \text{trace}(S\Theta) + \lambda \sum_{j \neq k} |\theta_{j,k}| + < \Gamma, \Theta > \rightarrow \min$$

1. $\Theta \succeq 0$
2. $\Gamma \succeq 0, \Theta^{-1} - S - \lambda \Psi + < \Gamma, \partial \Theta >= 0$
3. $< \Gamma, \Theta >= 0$,

where $A \succeq 0$ indicates that A is nonnegative definite and $< A, B >$ denotes the inner product $\text{trace}(A^T B)$ of matrices A, B. Because $\Gamma, \Theta \succeq 0$, $< \Gamma, \Theta >= 0$ means $\text{trace}(\Theta \Gamma) = \text{trace}(\Gamma^{1/2} \Theta \Gamma^{1/2}) = 0$, which further means that $\Gamma = 0$ because $\Gamma^{1/2} \Theta \Gamma^{1/2} \succeq 0$. Thus, we obtain (5.7).

Next, we obtain the solution Θ of (5.12). Let W be the inverse matrix of Θ. Decomposing the matrix $\Psi, \Theta, S, W \in \mathbb{R}^{p \times p}$, with the size of the upper-left elements being $(p - 1) \times (p - 1)$, we write

$$\Psi = \begin{bmatrix} \Psi_{1,1} & \psi_{1,2} \\ \psi_{2,1} & \psi_{2,2} \end{bmatrix}, \quad S = \begin{bmatrix} S_{1,1} & s_{1,2} \\ s_{2,1} & s_{2,2} \end{bmatrix}, \quad \Theta = \begin{bmatrix} \Theta_{1,1} & \theta_{1,2} \\ \theta_{2,1} & \theta_{2,2} \end{bmatrix}$$

$$\begin{bmatrix} W_{1,1} & w_{1,2} \\ w_{2,1} & w_{2,2} \end{bmatrix} \begin{bmatrix} \Theta_{1,1} & \theta_{1,2} \\ \theta_{2,1} & \theta_{2,2} \end{bmatrix} = \begin{bmatrix} I_{p-1} & 0 \\ 0 & 1 \end{bmatrix}, \tag{5.13}$$

where we assume $\theta_{2,2} > 0$. If Σ is positive definite, the eigenvalues of Θ are the inverse of those of Σ, and Θ is positive definite. Thus, if we multiply the vector with the first $p - 1$ and the p-th zero and one from the front and end, then the value (quadratic form) should be positive.

Then, from the upper-right part of both sides of (5.12), we have

$$w_{1,2} - s_{1,2} - \lambda \psi_{1,2} = 0 \tag{5.14}$$

where the upper-right part of Θ^{-1} is $w_{1,2}$. From the upper-right part of both sides in (5.13), we have

$$W_{1,1} \theta_{1,2} + w_{1,2} \theta_{2,2} = 0 . \tag{5.15}$$

If we put $\beta = \begin{bmatrix} \beta_1 \\ \vdots \\ \beta_{p-1} \end{bmatrix} := -\dfrac{\theta_{1,2}}{\theta_{2,2}}$, then from (5.14), (5.15), we have the following

equations:

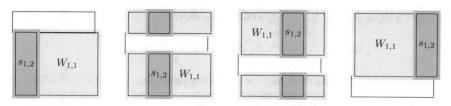

Fig. 5.3 The core part of graphical Lasso (Friedman et al., 2008)[12]. Update $W_{1,1}$, $w_{1,2}$, where the yellow and green elements are $W_{1,1}$ and $w_{1,2}$, respectively. The same value as that of $w_{1,2}$ is stored in $w_{2,1}$. In this case ($p = 4$), we repeat the four steps. The diagonal elements remain the same as those of S

$$W_{1,1}\beta - s_{1,2} + \lambda\phi_{1,2} = 0 \tag{5.16}$$

$$w_{1,2} = W_{1,1}\beta, \tag{5.17}$$

where the j-th element of $\phi_{1,2} \in \mathbb{R}^{p-1}$ is $\begin{cases} 1, & \beta_j > 0 \\ [-1, 1], & \beta_j = 0 \\ -1, & \beta_j < 0 \end{cases}$. In fact, each element

of $\psi_{1,2} \in \mathbb{R}^{p-1}$ is determined by the sign (positive/negative/zero) of $\theta_{1,2}$. Moreover, the values included in $[-1, 1]$ are included in it even if we flip the sign $[-1, 1]$. From $\theta_{2,2} > 0$, the signs of the elements in β and $\theta_{1,2}$ are opposite, and we have $\psi_{1,2} = -\phi_{1,2}$.

We find the solution of (5.12) as follows: First, we find the solution β of (5.16), substitute it into (5.17), and obtain $w_{1,2}$, which is the same as $w_{2,1}$ due to symmetry. We repeat the process, changing the positions of $W_{1,1}$, $w_{1,2}$. For $j = 1, \ldots, p$, $W_{1,1}$ is regarded as W, except in the j-th row and j-th column; $w_{1,2}$ is the W of the j-th column, except for the j-th element (Fig. 5.3).

If we input $A := W_{1,1} \in \mathbb{R}^{(p-1)\times(p-1)}$, $b = s_{1,2} \in \mathbb{R}^{p-1}$, and $c_j := b_j - \sum_{k\neq j} a_{j,k}\beta_k$, then each β_j that satisfies (5.16) is computed as

$$\beta_j = \begin{cases} \dfrac{c_j - \lambda}{a_{j,j}}, & c_j > \lambda \\ 0, & -\lambda < c_j < \lambda \\ \dfrac{c_j + \lambda}{a_{j,j}}, & c_j < -\lambda \end{cases}. \tag{5.18}$$

In the last stage, for $j = 1, \ldots, p$, if we take the following one cycle, we obtain the estimate of Θ

$$\theta_{2,2} = [w_{2,2} - w_{1,2}\beta]^{-1} \tag{5.19}$$

$$\theta_{1,2} = -\beta\theta_{2,2}, \tag{5.20}$$

where (5.19) is derived from (5.13) and $\beta = -\theta_{1,2}/\theta_{2,2}$.

$$\theta_{2,2}(w_{2,2} - w_{1,2}\beta) = \theta_{2,2}w_{2,2} - \theta_{2,2}w_{1,2}(-\theta_{1,2}/\theta_{2,2}) = 1$$

For example, we may construct the following procedure.

```
inner.prod = function(x, y) return(sum(x * y))   ## already appeared
soft.th = function(lambda, x) return(sign(x) * pmax(abs(x) - lambda,
    0))  ## already appeared
graph.lasso = function(s, lambda = 0) {
  W = s; p = ncol(s); beta = matrix(0, nrow = p - 1, ncol = p)
  beta.out = beta; eps.out = 1
  while (eps.out > 0.01) {
    for (j in 1:p) {
      a = W[-j, -j]; b = s[-j, j]
      beta.in = beta[, j]; eps.in = 1
      while (eps.in > 0.01) {
        for (h in 1:(p - 1)) {
          cc = b[h] - inner.prod(a[h, -h], beta[-h, j])
          beta[h, j] = soft.th(lambda, cc) / a[h, h]
        }
        eps.in = max(beta[, j] - beta.in); beta.in = beta[, j]
      }
      W[-j, j] = W[-j, -j] %*% beta[, j]
    }
    eps.out = max(beta - beta.out); beta.out = beta
  }
  theta = matrix(nrow = p, ncol = p)
  for (j in 1:p) {
    theta[j, j] = 1 / (W[j, j] - W[j, -j] %*% beta[, j])
    theta[-j, j] = -beta[, j] * theta[j, j]
  }
  return(theta)
}
```

Example 47 We construct an undirected graph by connecting s, t such that $\theta_{s,t} \neq 0$ is an edge. To this end, we generate data consisting of $N = 100$ tuples of $p = 5$ variables based on $\Theta = (\theta_{s,t})$, the values of which are known a priori.

```
library(MultiRNG)
Theta = matrix(c(   2,   0.6,     0,    0,  0.5,  0.6,    2, -0.4,  0.3,
         0,     0, -0.4,    2,
                  -0.2,    0,    0,  0.3, -0.2,    2, -0.2,  0.5,    0,
       0, -0.2,    2),
             nrow = 5)
Sigma = solve(Theta)
meanvec = rep(0, 5)
dat = draw.d.variate.normal(no.row = 20, d = 5, mean.vec = meanvec,
     cov.mat = Sigma)
# average: mean.vec, cov matric: cov.mat, sample #: no.row, variable
     #: d
s = t(dat) %*% dat / nrow(dat)
```

For the data, we detect whether an edge exists using the R language function graph.lasso, which recognizes whether each element of the matrix Theta is zero.

```
Theta
```

```
      [,1]  [,2]  [,3]  [,4]  [,5]
[1,]  2.0   0.6   0.0   0.0   0.5
[2,]  0.6   2.0  -0.4   0.3   0.0
[3,]  0.0  -0.4   2.0  -0.2   0.0
[4,]  0.0   0.3  -0.2   2.0  -0.2
[5,]  0.5   0.0   0.0  -0.2   2.0
```

```
graph.lasso(s)
```

```
            [,1]         [,2]        [,3]         [,4]        [,5]
[1,]  2.1854523   0.49532753  -0.1247222  -0.04480949   0.46042784
[2,]  0.4962869   1.75439746  -0.2578557   0.17696543  -0.04818041
[3,] -0.1258535  -0.25806533   2.0279864  -0.47076829   0.13461843
[4,] -0.0435247   0.17742249  -0.4710045   2.22329991  -0.45533867
[5,]  0.4595971  -0.04867342   0.1350065  -0.45560256   2.20517785
```

```
graph.lasso(s, lambda = 0.015)
```

```
            [,1]        [,2]         [,3]        [,4]        [,5]
[1,]  2.12573980   0.4372959  -0.06516687   0.0000000   0.39343910
[2,]  0.43823782   1.7088250  -0.17041729   0.1011818   0.00000000
[3,] -0.06538542  -0.1708783   1.97559922  -0.3728077   0.04907338
[4,]  0.00000000   0.1010347  -0.37528158   2.1517477  -0.34531269
[5,]  0.39473905   0.0000000   0.04930174  -0.3453602   2.14001693
```

```
graph.lasso(s, lambda = 0.03)
```

```
            [,1]         [,2]          [,3]         [,4]        [,5]
[1,]  2.0764031   0.37682657  -0.0002649865   0.00000000   0.3254736
[2,]  0.3769137   1.67661814  -0.0963712162   0.03264331   0.0000000
[3,]  0.0000000  -0.09711355   1.9435147152  -0.29372726   0.0000000
[4,]  0.0000000   0.03199325  -0.2935709011   2.10377862  -0.2659885
[5,]  0.3271177   0.00000000   0.0000000000  -0.26606666   2.0965524
```

```
graph.lasso(s, lambda = 0.05)
```

	[,1]	[,2]	[,3]	[,4]	[,5]
[1,]	2.0259641	0.30684893	0.00000000	0.0000000	0.2421536
[2,]	0.3068663	1.64928815	-0.03096349	0.0000000	0.0000000
[3,]	0.0000000	-0.03157345	1.91889676	-0.2112410	0.0000000
[4,]	0.0000000	0.00000000	-0.21143915	2.0642104	-0.1808793
[5,]	0.2423494	0.00000000	0.00000000	-0.1809381	2.0549861

For the data generated above, we draw the undirected graphs $\lambda = 0, 0.015, 0.03$, 0.05 in Fig. 5.4. We observe that the larger the value of λ is, the greater the number of edges in the graph.

In the actual data processing w.r.t. graphical Lasso, the `glasso` R package is often used (the option `rho` is used to specify λ). In this book, we construct the function `graphical.lasso`, but the process is not optimized for finer details. In the rest of this section, we use `glasso`.

Example 48 We can execute the same procedure as in Example 47 by `glasso`.

```
library(glasso)
solve(s); glasso(s, rho = 0); glasso(s, rho = 0.015)
glasso(s, rho = 0.030); glasso(s, rho = 0.045)
```

When we use a graph of the numerical data output of the R language, for both directed and undirected graphs, the `igraph` R package is used. Using the package, we draw an undirected graph such that (i, j) are connected if and only if $a_{i,j} \neq 0$ from a symmetric matrix $A = (a_{i,j})$ of size p. Moreover, we use it to draw an undirected graph from a precision matrix Θ. For example, we can construct the following function `adj`.

```
library(igraph)
adj = function(mat) {
  p = ncol(mat); ad = matrix(0, nrow = p, ncol = p)
  for (i in 1:(p - 1)) for (j in (i + 1):p) {
    if (mat[i, j] == 0) ad[i, j] = 0 else ad[i, j] = 1
  }
  g = graph.adjacency(ad, mode = "undirected")
  plot(g)
}
```

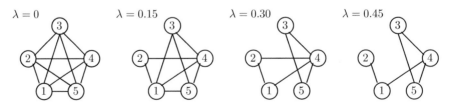

Fig. 5.4 In Example 47, we fix an undirected graph and generate data based on the graph. From the data, we estimate the undirected graph from the samples. We observe that the larger the value of λ is, the greater the number of edges in the graph

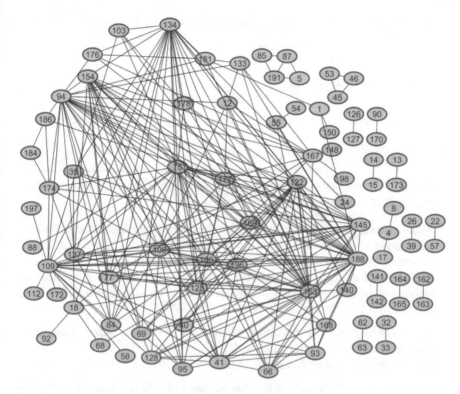

Fig. 5.5 We draw the output of `glasso` w.r.t. the first 200 genes for $\lambda = 0.75$. For display, we used a drawing application (Cytoscape) rather than `igraph`

However, it is not easy to draw a graph with hundreds of vertices in a display at once. In particular, in the `igraph` package, the vertices and edges overlap even if the vertex size is shrunk. Thus, it may be appropriate to use the R procedure only for obtaining the adjacency matrix and to use another drawing software program for displaying the graph.[2] Using this, application software is even better if the display is used for paper submission, an important presentation, etc. It may be appropriate to say that the task of structure estimation of a graphical model is to obtain the mathematical description of the graph rather than to display the graph itself.

Example 49 (*Breast Cancer*) We execute graphical Lasso for the dataset breast-cancer.csv ($N = 250$, $p = 1000$). The purpose of the procedure is to find the relation among the gene expressions of breast cancer patients. When we execute the following program, we observe that the smaller the value of λ is, the longer the execution time. We draw the output of `glasso` w.r.t. the first 200 genes out of $p = 1000$ for $\lambda = 0.75$ (Fig. 5.5). The execution is carried out using the following code.

---

[2] Cytoscape (https://cytoscape.org/), etc.

```
1   library(glasso); library(igraph)
2   df = read.csv("breastcancer.csv")
3   w = matrix(nrow = 250, ncol = 1000)
4   for (i in 1:1000) w[, i] = as.numeric(df[[i]])
5   x = w; s = t(x) %*% x / 250
6   fit = glasso(s, rho = 0.75); sum(fit$wi == 0)
7   y = NULL; z = NULL
8   for (i in 1:999) for (j in (i + 1):1000) if (fit$wi[i, j] != 0) {y = c
        (y, i); z = c(z, j)}
9   edges = cbind(y, z)
10  write.csv(edges,"edges.csv")
```

The edges of the undirected graph are stored in edges.csv. We input the file to Cytoscape to obtain the output.

Although thus far, we have addressed the algorithm of graphical Lasso , we have not mentioned how to set the value of λ and the correctness of graphical Lasso for a large sample size N (missing/extra edges and parameter estimation).

Ravikumar et al. 2011 [23] presented a result that theoretically guarantees the correctness of the graphical Lasso procedure: there exist a parameter α and function f, g determined w.r.t. Θ such that when λ is $f(\alpha)/\sqrt{N}$,

1. the probability that the maximum element of $\Theta - \hat{\Theta}$ is upper bounded by $g(\alpha)/\sqrt{N}$ can be arbitrarily small and
2. the probability that the edge sets in Θ and $\hat{\Theta}$ coincide can be arbitrarily large

as the sample size N grows, where $\hat{\Theta}$ is the estimate of Θ that is obtained by graphical Lasso.

Although the paper suggests that λ should be $O(1/\sqrt{N})$, the parameter α is not known and cannot be estimated from the N samples, which means that the λ that guarantees the correctness cannot be obtained. In addition, the theory assumes a range of α, and we cannot know from the samples that the condition is met. Even if the λ value is known, the paper does not prove that the choice of λ is optimum. However, Ref. [23] claims that consistency can be obtained for a broad range of α for large N and that the parameter estimation improves with $O(1/\sqrt{N})$.

5.3 Estimation of the Graphical Model based on the Quasi-likelihood

If the response is continuous and discrete variables, we can infer the relevant covariates via linear and logistic regressions. Moreover, each covariate may be either discrete or continuous.

For a response X_i, if the covariates are X_k ($k \in \pi_i \subseteq \{1, \ldots, p\}\setminus\{i\}$), we consider the subset π_i to be the parent set of X_i. Even if we obtain the optimum parent set, we cannot maximize the (regularized) likelihood of the data. If the graph has a tree/forest structure, and if we seek the parent set from the root to the leaves, it may be possible

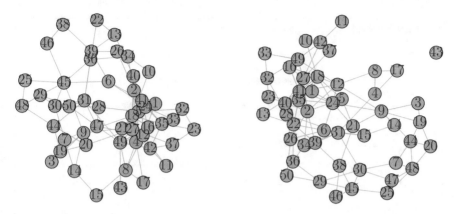

Fig. 5.6 The graphical model obtained via the quasi-likelihood method from the Breast Cancer Dataset. When all the variables are continuous (left, Example 50), and when all the variables are binary (right, Example 51)

to obtain the exact likelihood. However, we may wonder where the root should be, why we may artificially assume the structure, etc.

Thus, to approximately realize maximizing the likelihood, we obtain the parent set π_i ($i = 1, \ldots, p$) of each vertex (response) independently. We may choose one of the following:

1. $j \in \pi_i$ and $i \in \pi_j \Longrightarrow$ Connect i, j as an edge (AND rule)
2. $j \in \pi_i$ or $i \in \pi_j \Longrightarrow$ Connect i, j as an edge (OR rule),

where the choice of λ may be different for each response.

Although this quasi-likelihood method [20] may not be theoretical, it can be applied to general situations, in contrast to graphical Lasso:

1. the distribution over the p variables may not be Gaussian
2. some of the p variables may be discrete.

A demerit of the method is that when λ is small, particularly for large p, it is rather time-consuming compared with graphical Lasso.

If we realize it using the R language, we may use the `glmnet` package: the options `family = "gaussian"` (default), `family = "binomial"`, and `family = "multinomial"` are available for continuous, binary, and multiple variables, respectively.

Example 50 We generate a graph using the quasi-likelihood method via `glmnet` for the Breast Cancer Dataset. In the following, we examine the difference between the AND and OR rules. Although the data contain 1,000 variables, we execute the first $p = 50$ variables. Then, the difference cannot be seen between the AND and OR rules. For $\lambda = 0.1$, there are 243 edges out of the $p(p-1)/2 = 1,225$ candidates (Fig. 5.6, left).

```
1   library(glmnet)
2   df = read.csv("breastcancer.csv")
3   n = 250; p = 50; w = matrix(nrow = n, ncol = p)
4   for (i in 1:p) w[, i] = as.numeric(df[[i]])
5   x = w[, 1:p]; fm = rep("gaussian", p); lambda = 0.1
6   fit = list()
7   for (j in 1:p) fit[[j]] = glmnet(x[, -j], x[, j], family = fm[j],
        lambda = lambda)
8   ad = matrix(0, p, p)
9   for (i in 1:p) for (j in 1:(p - 1)) {
10    k = j
11    if (j >= i) k = j + 1
12    if (fit[[i]]$beta[j] != 0) ad[i, k] = 1 else ad[i, k] = 0
13  }
14  ## AND
15  for (i in 1:(p - 1)) for (j in (i + 1):p) {
16    if (ad[i, j] != ad[i, j]) {ad[i, j] = 0; ad[j, i] = 0}
17  }
18  u = NULL; v = NULL
19  for (i in 1:(p - 1)) for (j in (i + 1):p) {
20    if (ad[i, j] == 1) {u = c(u, i); v = c(v, j)}
21  }
22  u
```

```
1   [1]   1   1   1   1   1   1   1   1   1   1   1   1   1   2   2   2   2   2   2   2   2   2   2
2  [23]   2   2   3   3   3   3   3   3   3   3   3   3   4   4   4   4   4   4   4   5   5   5
3  . . . . . . . . . . . . . . . . . . . . . . . . . . . . . . . . . . . . . . . . . . . . . . . . . . . . . .
4 [199] 30 30 31 31 31 31 31 31 31 32 32 33 33 34 34 34 35 35 36 36 36 37
5 [221] 37 37 38 38 38 38 38 40 40 42 42 43 43 43 44 44 45 45 47 47 47 49
```

```
1   v
```

```
1   [1]   2  12  13  18  22  23  24  28  41  46  48  50   4   5  10  16  18  21  23  26  27  34
2  [23] 37  43   9  10  15  19  20  21  24  29  38  50   5   8  17  18  21  23  46   9  20  21
3  . . . . . . . . . . . . . . . . . . . . . . . . . . . . . . . . . . . . . . . . . . . . . . . . . . . . . .
4 [199] 48  50  34  36  39  44  47  49  50  33  45  37  49  35  36  45  41  42  38  39  45  38
5 [221] 42  46  39  45  46  49  50  41  42  49  50  47  49  50  48  50  46  50  48  49  50  50
```

```
1   adj(ad)
2   ## OR
3   for (i in 1:(p - 1)) for (j in (i + 1):p) {
4     if (ad[i, j] != ad[i, j]) {ad[i, j] = 1; ad[j, i] = 1}
5   }
6   adj(ad)
```

Example 51 We construct an undirected graph from N samples of p binary variables (the sign $\pm$ of the values in breastcancer.csv).

```
1  library(glmnet)
2  df = read.csv("breastcancer.csv")
3  w = matrix(nrow = 250, ncol = 1000); for (i in 1:1000) w[, i] = as.
       numeric(df[[i]])
4  w = (sign(w) + 1) / 2  ## transforming it to binary
5  p = 50; x = w[, 1:p]; fm = rep("binomial", p); lambda = 0.15
6  fit = list()
7  for (j in 1:p) fit[[j]] = glmnet(x[, -j], x[, j], family = fm[j],
       lambda = lambda)
8  ad = matrix(0, nrow = p, ncol = p)
9  for (i in 1:p) for (j in 1:(p - 1)) {
10   k = j
11   if (j >= i) k = j + 1
12   if (fit[[i]]$beta[j] != 0) ad[i, k] = 1 else ad[i, k] = 0
13 }
14 for (i in 1:(p - 1)) for (j in (i + 1):p) {
15   if (ad[i, j] != ad[i, j]) {ad[i, j] = 0; ad[j, i] = 0}
16 }
17 sum(ad); adj(ad)
```

We may mix continuous and discrete variables by setting one of the options
"gaussian","binomial","multinomial" for each response in the array
fm of Examples 50 and 51.

5.4 Joint Graphical Lasso

Thus far, we have generated undirected graphs from the data $X \in \mathbb{R}^{N \times p}$. Next, we
consider generating a graph from each of $X_1 \in \mathbb{R}^{N_1 \times p}$ and $X_2 \in \mathbb{R}^{N_2 \times p}$ ($N_1 + N_2 =
N$) (joint graphical Lasso). We may assume a supervised learning with $y \in \{1, 2\}^N$
such that X_1, X_2 are generated from the rows i of X with $y_i = 1$ and $y_i = 2$. The
idea is to utilize the N samples when the two graphs share similar properties: if we
generate two graphs from X_1 and X_2 separately, the estimates will be poorer than
those obtained by applying the joint graphical Lasso. For example, suppose that we
generate two gene regulatory networks for the case and control expression data of
p genes. In this case, the two graphs would be similar, although some edges may be
different.

The joint graphical Lasso (JGL; Danaher-Witten, 2014) [8] extends the maximization of (5.10) to that of

$$\sum_{i=1}^{K} N_k \log \det \Theta_k - \text{trace}(S\Theta_k) - P(\Theta_1, \ldots, \Theta_K), \qquad (5.21)$$

for $y \in \{1, 2, \ldots, K\}$ ($K \geq 2$).

For the $P(\Theta_1, \ldots, \Theta_K)$ term, we choose one of the fused Lasso and group Lasso
options. The penalties of them are, respectively,

$$P(\Theta_1, \ldots, \Theta_K) := \lambda_1 \sum_k \sum_{i \neq j} |\theta_{i,j}^{(k)}| + \lambda_2 \sum_{k<k'} \sum_{i,j} |\theta_{i,j}^{(k)} - \theta_{i,j}^{(k')}| \qquad (5.22)$$

$$P(\Theta_1, \ldots, \Theta_K) := \lambda_1 \sum_k \sum_{i \neq j} |\theta_{i,j}^{(k)}| + \lambda_2 \sum_{i \neq j} \sqrt{\sum_k \theta_{i,j}^{(k)2}}, \qquad (5.23)$$

where the index of $\sum$ ranges over $k = 1, \ldots, K$, $i, j = 1, \ldots, p$. The Eqs. (5.22), (5.23) share the first term that makes the small absolute value terms zero, similar to the Lasso procedures that we have seen.

In JGL, to obtain the solution of (5.21), we apply the ADMM: We define the extended Lagrange as

$$L_\rho(\Theta, Z, U) := -\sum_{k=1}^K N_k \{\log \det \Theta_k - \text{trace}(S\Theta_k)\} + P(Z_1, \ldots, Z_K)$$

$$+ \rho \sum_{k=1}^K \langle U_k, \Theta_k - Z_k \rangle + \frac{\rho}{2} \sum_{k=1}^K \|\Theta_k - Z_k\|_F^2$$

and repeat the following three steps:

i. $\Theta^{(t)} \leftarrow \text{argmin}_\Theta L_\rho(\Theta, Z^{(t-1)}, U^{(t-1)})$
ii. $Z^{(t)} \leftarrow \text{argmin}_Z L_\rho(\Theta^{(t)}, Z, U^{(t-1)})$
iii. $U^{(t)} \leftarrow U^{(t-1)} + \rho(\Theta^{(t)} - Z^{(t)})$.

More precisely, setting $\rho > 0$ and letting Θ_k and Z_k, U_k be the unit and zero matrices for $k = 1, \ldots, K$, we repeat the above three steps. For the first step, if differentiating $L_\rho(\Theta, Z, U)$ by Θ, then from a similar discussion as for graphical Lasso, we have

$$-N_k(\Theta_k^{-1} - S_k) + \rho(\Theta_k - Z_k + U_k) = 0.$$

If we decompose both sides (symmetric matrix) of

$$\Theta_k^{-1} - \frac{\rho}{N_k}\Theta_k = S_k - \rho\frac{Z_k}{N_k} + \rho\frac{U_k}{N_k}$$

as VDV^T (D: a diagonal matrix), since Θ_k and VDV^T share the same eigenvector, we can write $\Theta_k = V\tilde{D}V^T$, $\Theta_k^{-1} = V\tilde{D}^{-1}V^T$. Thus, for the diagonal elements, we have

$$\frac{1}{\tilde{D}_{j,j}} - \frac{\rho}{N_k}\tilde{D}_{j,j} = D_{j,j},$$

i.e.,

$$\tilde{D}_{j,j} = \frac{N_k}{2\rho}\left(-D_{j,j} + \sqrt{D_{j,j}^2 + 4\rho/N_k}\right).$$

Therefore, the optimum Θ_k can be written as $V\tilde{D}V^T$ using this $\tilde{D}$.

The second step executes fused Lasso among different classes k. For $K = 2$, genlasso and other fused Lasso procedures do not work; thus, we set up the function b.fused. If the difference between y_1 and y_2 is at most 2λ, then $\theta_1 = \theta_2 = (y_1 + y_2)/2$. Otherwise, we subtract λ from the larger of y_1, y_2 and add λ to the smaller one. The procedure to obtain λ_1 from λ_2 is due to the sparse fused Lasso discussed in Sect. 4.2.

We construct the JGL for fused Lasso as follows.

```
# genlasso works only when the size is at least three
b.fused = function(y, lambda) {
  if (y[1] > y[2] + 2 * lambda) {a = y[1] - lambda; b = y[2] + lambda}
  else if (y[1] < y[2] - 2 * lambda) {a = y[1] + lambda; b = y[2] - lambda}
  else {a = (y[1] + y[2]) / 2; b = a}
  return(c(a, b))
}
# fused Lasso that compares not only the adjacency terms but also all
    adjacency values
fused = function(y, lambda.1, lambda.2) {
  K = length(y)
  if (K == 1) theta = y
  else if (K == 2) theta = b.fused(y, lambda.2)
  else {
    L = K * (K - 1) / 2; D = matrix(0, nrow = L, ncol = K)
    k = 0
    for (i in 1:(K - 1)) for (j in (i + 1):K) {
      k = k + 1; D[k, i] = 1; D[k, j] = -1
    }
    out = genlasso(y, D = D)
    theta = coef(out, lambda = lambda.2)
  }
  theta = soft.th(lambda.1, theta)
  return(theta)
}
# Joint Graphical Lasso
jgl = function(X, lambda.1, lambda.2) {  # X is given as a list
  K = length(X); p = ncol(X[[1]]); n = array(dim = K); S = list()
  for (k in 1:K) {n[k] = nrow(X[[k]]); S[[k]] = t(X[[k]]) %*% X[[k]] / n[k]}
  rho = 1; lambda.1 = lambda.1 / rho; lambda.2 = lambda.2 / rho
  Theta = list(); for (k in 1:K) Theta[[k]] = diag(p)
  Theta.old = list(); for (k in 1:K) Theta.old[[k]] = diag(rnorm(p))
  U = list(); for (k in 1:K) U[[k]] = matrix(0, nrow = p, ncol = p)
  Z = list(); for (k in 1:K) Z[[k]] = matrix(0, nrow = p, ncol = p)
  epsilon = 0; epsilon.old = 1
  while (abs((epsilon - epsilon.old) / epsilon.old) > 0.0001) {
    Theta.old = Theta;  epsilon.old = epsilon
    ## Update (i)
    for (k in 1:K) {
      mat = S[[k]] - rho * Z[[k]] / n[k] + rho * U[[k]] / n[k]
      svd.mat = svd(mat)
      V = svd.mat$v
      D = svd.mat$d
      DD = n[k] / (2 * rho) * (-D + sqrt(D ^ 2 + 4 * rho / n[k]))
      Theta[[k]] = V %*% diag(DD) %*% t(V)
    }
    ## Update (ii)
    for (i in 1:p) for (j in 1:p) {
      A = NULL; for (k in 1:K) A = c(A, Theta[[k]][i, j] + U[[k]][i, j])
```

```
49      if (i == j) B = fused(A, 0, lambda.2) else B = fused(A, lambda.1,
        lambda.2)
50        for (k in 1:K) Z[[k]][i, j] = B[k]
51      }
52      ## Update (iii)
53      for (k in 1:K) U[[k]] = U[[k]] + Theta[[k]] - Z[[k]]
54      ## Test Convergence
55      epsilon = 0
56      for (k in 1:K) {
57        epsilon.new = max(abs(Theta[[k]] - Theta.old[[k]]))
58        if (epsilon.new > epsilon) epsilon = epsilon.new
59      }
60    }
61    return(Z)
62  }
```

For the JGL for the group Lasso, we may replace the update (ii). Let $A_k[i, j] := \Theta_k[i, j] + U_k[i, j]$. Then, no update is required for $i = j$, and

$$Z_k[i, j] = \mathcal{S}_{\lambda_1/\rho}(A_k[i, j]) \left(1 - \frac{\lambda_2}{\rho\sqrt{\sum_{k=1}^{K} \mathcal{S}_{\lambda_1/\rho}(A_k[i, j])^2}} \right)_+$$

for $i \neq j$.

We construct the following code, in which no functions b.fused and genlasso are required.

```
1  ## Replace the update (ii) by the following
2  for (i in 1:p) for (j in 1:p) {
3    A = NULL; for (k in 1:K) A = c(A, Theta[[k]][i, j] + U[[k]][i, j])
4    if (i == j) B = A
5    else {B = soft.th(lambda.1 / rho,A) *
6      max(1 - lambda.2 / rho / sqrt(norm(soft.th(lambda.1 / rho, A), "2"
        ) ^ 2), 0)}
7    for (k in 1:K) Z[[k]][i, j] = B[k]
8  }
```

Example 52 For $K = 2$, adding noise to the data $X \in \mathbb{R}^{N \times p}$, we generate similar data $X' \in \mathbb{R}^{N \times p}$. For the data X, X' and the parameters $(\lambda_1, \lambda) = (10, 0.05)$, $(10, 0.10)$, $(3, 0.03)$, we execute the fused Lasso JGL with the following code.

```
1  ## Data Generation and Execution
2  p = 10; K = 2; N = 100; n = array(dim = K); for (k in 1:K) n[k] = N /
      K
3  X = list(); X[[1]] = matrix(rnorm(n[k] * p), ncol = p)
4  for (k in 2:K) X[[k]] = X[[k - 1]] + matrix(rnorm(n[k] * p) * 0.1,
      ncol = p)
5  ## Change the lambda.1,lambda.2 values to execute
6  Theta = jgl(X, 3, 0.01)
7  par(mfrow = c(1, 2)); adj(Theta[[1]]); adj(Theta[[2]])
```

We observe that the larger the value of λ_1 is, the more sparse the graphs. On the other hand, the larger the value of λ_2 is, the more similar the graphs (Fig. 5.7).

(a) $\lambda_1 = 10, \lambda_2 = 0.05$

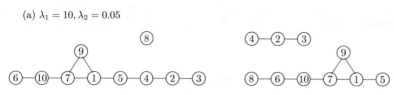

(b) $\lambda_1 = 10, \lambda_2 = 0.10$

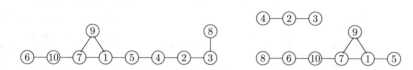

(c) $\lambda_1 = 3, \lambda_2 = 0.03$

	1	2	3	4	5	6	7	8	9	10
1		0	1	0	1	0	1	1	1	0
2			1	1	1	1	0	0	1	0
3				0	0	0	1	0	0	1
4					1	1	1	0	0	1
5						0	1	1	1	1
6							0	1	1	0
7								0	1	1
8									1	1
9										1

	1	2	3	4	5	6	7	8	9	10
1		1	0	0	1	1	1	1	1	0
2			0	1	1	1	0	0	1	0
3				0	0	0	1	0	1	1
4					1	1	1	0	0	0
5						1	0	1	0	1
6							0	0	1	1
7								0	1	1
8									1	1
9										1

Fig. 5.7 Execution of Example 52 (fused Lasso, $K = 2$) with $(\lambda_1, \lambda) = (10, 0.05), (10, 0.10), (3, 0.03)$. We observe that the larger the value of λ_1 is, the more sparse the graphs, and that the larger the value of λ_2 is, the more similar the graphs

Fig. 5.8 Execution of Example 52 (group Lasso, $K = 2$) with $(\lambda_1, \lambda) = (10, 0.01)$

We execute the group Lasso JGL (Fig. 5.8). From the example, little difference can be observed from the fused Lasso JGL.

Appendix: Proof of Propositions

Proposition 15 *(Lauritzen, 1996 [18])*

$$\Theta_{AB} = \Theta_{BA}^T = 0 \implies \det \Sigma_{A \cup C} \det \Sigma_{B \cup C} = \det \Sigma_{A \cup B \cup C} \det \Sigma_C$$

Proof: Applying *Proof* $a = \begin{bmatrix} \Theta_{AA} & \Theta_{AC} \\ \Theta_{CA} & \Theta_{CC} \end{bmatrix}$, $b = \begin{bmatrix} 0 \\ \Theta_{CB} \end{bmatrix}$, $c = \begin{bmatrix} 0 & \Theta_{BC} \end{bmatrix}$, $d = \Theta_{BB}$,

$$e = a - bd^{-1}c = \begin{bmatrix} \Theta_{AA} & \Theta_{AC} \\ \Theta_{CA} & \Theta_{CC} \end{bmatrix} - \begin{bmatrix} 0 \\ \Theta_{CB} \end{bmatrix} \Theta_{BB}^{-1} \begin{bmatrix} 0 & \Theta_{BC} \end{bmatrix}$$
$$= \begin{bmatrix} \Theta_{AA} & \Theta_{AC} \\ \Theta_{CA} & \Theta_{CC} - \Theta_{CB}(\Theta_{BB})^{-1}\Theta_{BC} \end{bmatrix},$$

in

$$\begin{bmatrix} a & b \\ c & d \end{bmatrix}^{-1} = \begin{bmatrix} e^{-1} & -e^{-1}bd^{-1} \\ -ed^{-1}c & d^{-1} + d^{-1}ce^{-1}bd^{-1} \end{bmatrix}, \tag{5.24}$$

we have

$$\begin{bmatrix} \Sigma_{A \cup C} & * \\ * & * \end{bmatrix} = \Theta^{-1} = \begin{bmatrix} \Theta_{AA} & \Theta_{AC} & 0 \\ \Theta_{CA} & \Theta_{CC} & \Theta_{CB} \\ 0 & \Theta_{BC} & \Theta_{BB} \end{bmatrix}^{-1}$$
$$= \begin{bmatrix} \begin{bmatrix} \Theta_{AA} & \Theta_{AC} \\ \Theta_{CA} & \Theta_{CC} - \Theta_{CB}(\Theta_{BB})^{-1}\Theta_{BC} \end{bmatrix}^{-1} & * \\ * & * \end{bmatrix},$$

which means

$$(\Sigma_{A \cup C})^{-1} = \begin{bmatrix} \Theta_{AA} & \Theta_{AC} \\ \Theta_{CA} & \Theta_{CC} - \Theta_{CB}(\Theta_{BB})^{-1}\Theta_{BC} \end{bmatrix}, \tag{5.25}$$

where (5.24) is due to

$$\begin{bmatrix} a & b \\ c & d \end{bmatrix} \begin{bmatrix} e^{-1} & -e^{-1}bd^{-1} \\ -d^{-1}ce^{-1} & d^{-1}(I + ce^{-1}bd^{-1}) \end{bmatrix}$$
$$= \begin{bmatrix} ae^{-1} - bd^{-1}ce^{-1} & -ae^{-1}bd^{-1} + bd^{-1}(I + ce^{-1}bd^{-1}) \\ ce^{-1} - ce^{-1} & -ce^{-1}bd^{-1} + I + ce^{-1}bd^{-1} \end{bmatrix} = \begin{bmatrix} I & 0 \\ 0 & I \end{bmatrix}.$$

Similarly, we have

$$(\Sigma_C)^{-1} = \Theta_{CC} - \begin{bmatrix} \Theta_{CA} & \Theta_{CB} \end{bmatrix} \begin{bmatrix} \Theta_{AA} & 0 \\ 0 & \Theta_{BB} \end{bmatrix}^{-1} \begin{bmatrix} \Theta_{AC} \\ \Theta_{BC} \end{bmatrix}$$
$$= \Theta_{CC} - \Theta_{CA}(\Theta_{AA})^{-1}\Theta_{AC} - \Theta_{CB}(\Theta_{BB})^{-1}\Theta_{BC}.$$

Thus, we have

$$\det \begin{bmatrix} I & (\Theta_{AA})^{-1}\Theta_{AC} \\ \Theta_{CA} & \Theta_{CC} - \Theta_{CB}(\Theta_{BB})^{-1}\Theta_{BC} \end{bmatrix} = \det(\Sigma_C)^{-1} . \tag{5.26}$$

Furthermore, from the identity

$$\begin{bmatrix} \Theta_{AA}^{-1} & 0 \\ 0 & I \end{bmatrix} \begin{bmatrix} \Theta_{AA} & \Theta_{AC} \\ \Theta_{CA} & \Theta_{CC} - \Theta_{CB}(\Theta_{BB})^{-1}\Theta_{BC} \end{bmatrix} = \begin{bmatrix} I & (\Theta_{AA})^{-1}\Theta_{AC} \\ \Theta_{CA} & \Theta_{CC} - \Theta_{CB}(\Theta_{BB})^{-1}\Theta_{BC} \end{bmatrix}$$

and (5.25), (5.26), we have

$$\det \Sigma_{A\cup C} = \frac{\det(\Sigma_C)}{\det \Theta_{AA}} . \tag{5.27}$$

On the other hand, from

$$\det \begin{bmatrix} a & b \\ c & d \end{bmatrix} = \det \begin{bmatrix} a & b \\ c & d \end{bmatrix} \det \begin{bmatrix} I & O \\ -d^{-1}c & I \end{bmatrix} = \det(a - bd^{-1}c) \det d ,$$

we have

$$\det \Theta = \det \begin{bmatrix} \begin{bmatrix} \Theta_{AA} \\ 0 \end{bmatrix} & \begin{bmatrix} 0 \\ \Theta_{CA} \end{bmatrix} \\ \begin{bmatrix} 0 & \Theta_{CA} \end{bmatrix} & \Theta_{B\cup C} \end{bmatrix} = \det \begin{bmatrix} \Theta_{B\cup C} & \begin{bmatrix} 0 \\ \Theta_{CA} \end{bmatrix} \\ \begin{bmatrix} 0 & \Theta_{CA} \end{bmatrix} & \Theta_{AA} \end{bmatrix}$$

$$= \det \left\{ \Theta_{B\cup C} - \begin{bmatrix} 0 \\ \Theta_{CA} \end{bmatrix} \Theta_{AA} \begin{bmatrix} 0 & \Theta_{CA} \end{bmatrix} \right\} \cdot \det \Theta_{AA} = \frac{\det \Theta_{AA}}{\det \Sigma_{B\cup C}} . \tag{5.28}$$

By multiplying both sides of (5.27), (5.28), we have

$$\det \Theta = \frac{\det \Sigma_C}{\det \Sigma_{A\cup C} \det \Sigma_{B\cup C}} .$$

From $\det \Theta = (\det \Sigma)^{-1}$, this proves the proposition. □

Proposition 17 *(Pearl and Paz, 1985)* Suppose that we construct an undirected graph (V, E) such that $\{i, j\} \in E \Longleftrightarrow X_i \perp\!\!\!\perp_P X_j \mid X_S$ for $i, j \in V$ and $S \subseteq V$. Then, $X_A \perp\!\!\!\perp_P X_B \mid X_C \Longleftrightarrow A \perp\!\!\!\perp_E B \mid C$ for all disjoint subsets A, B, C of V if and only if the probability P satisfies the following conditions:

$$X_A \perp\!\!\!\perp_P X_B \mid X_C \Longleftrightarrow X_B \perp\!\!\!\perp_P X_A \mid X_C \tag{5.4}$$

$$X_A \perp\!\!\!\perp_P (X_B \cup X_D) \mid X_C \Longrightarrow X_A \perp\!\!\!\perp_P X_B \mid X_C, X_A \perp\!\!\!\perp_P X_D \mid X_C \tag{5.5}$$

$$X_A \perp\!\!\!\perp_P (X_C \cup X_D) \mid X_B, X_A \perp\!\!\!\perp_E X_D \mid (X_B \cup X_C) \Longrightarrow X_A \perp\!\!\!\perp_E (X_B \cup X_D) \mid X_C \tag{5.6}$$

$$X_A \perp\!\!\!\perp_P X_B \mid X_C \Longrightarrow X_A \perp\!\!\!\perp_P X_B \mid (X_C \cup X_D) \qquad (5.7)$$

$$X_A \perp\!\!\!\perp_P X_B \mid X_C \Longrightarrow X_A \perp\!\!\!\perp_P X_k \mid X_C \text{ or } X_k \perp\!\!\!\perp_E X_B \mid X_C \qquad (5.8)$$

for $k \notin A \cup B \cup C$, where A, B, C, D are disjoint subsets of V for each equation.

Proof: Note that the first four conditions imply

$$X_A \perp\!\!\!\perp_P X_B \mid X_C \Longleftrightarrow X_i \perp\!\!\!\perp_P X_j \mid X_C \; i \in A, j \in B \qquad (5.29)$$

because the converse of (5.5) subsequently holds:

$$X_A \perp\!\!\!\perp_P X_B \mid X_C , \; X_A \perp\!\!\!\perp_P X_D \mid X_C \Longrightarrow X_A \perp\!\!\!\perp_P X_B \mid (X_C \cup X_D), X_A \perp\!\!\!\perp_P X_D \mid (X_C \cup X_B)$$
$$\Longrightarrow X_A \perp\!\!\!\perp_P (X_B \cup X_D) \mid X_C ,$$

where (5.6) and (5.7) have been used. Thus, it is sufficient to show

$$X_i \perp\!\!\!\perp_P X_j \mid X_S \Longleftrightarrow i \perp\!\!\!\perp_E j \mid S$$

for $i, j \in U$ and $S \subseteq U$. From the construction of the undirected graph, $\Longleftarrow$ is obvious.

For $\Longrightarrow$, if $|S| = p - 2$, the theorem holds because (V, E) was constructed so. Suppose that the theorem holds for $|S| = r \leq p - 2$ and that the size of S' is $|S'| = r - 1$, and let $k \notin S' \cup \{i, j\}$.

1. From (5.7), we have $i \perp\!\!\!\perp_P j \mid (S' \cup k)$.
2. From (5.8), we have either $i \perp\!\!\!\perp_P k \mid S'$ or $j \perp\!\!\!\perp_P k \mid S'$, which means that we have $i \perp\!\!\!\perp_P k \mid (S' \cup j)$ from (5.7).
3. Since the size of $S' \cup j$ and $S' \cup k$ is r, by an induction hypothesis, from the first and second, we have $i \perp\!\!\!\perp_E j \mid (S' \cup k)$ and $i \perp\!\!\!\perp_E k \mid (S' \cup j)$.
4. From (5.6), from the third item, we have $i \perp\!\!\!\perp_E j \mid S'$,

which establish $\Longrightarrow$.   □

Exercises 62–75

62. In what graph among (a) through (f) do the red vertices separate the blue and green vertices?

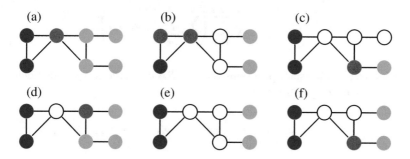

63. Let X, Y be binary variables that take zeros and ones and are independent, and let Z be the residue of the sum when divided by two. Show that X, Y are not conditionally independent given Z. Note that the probabilities p and q of $X = 1$ and $Y = 1$ are not necessarily 0 and 5 and that we do not assume $p = q$.

64. For the precision matrix $\Theta = (\theta_{i,j})_{i,j=1,2,3}$, with $\theta_{12} = \theta_{21} = 0$, show

$$\det \Sigma_{\{1,2,3\}} \det \Sigma_{\{3\}} = \det \Sigma_{\{1,3\}} \det \Sigma_{\{2,3\}} \,,$$

where by Σ_S, we mean the submatrix of Σ that consists of rows and columns with indices $S \subseteq \{1, 2, 3\}$.

65. Suppose that we express the probability density function of the p Gaussian variables with mean zero and precision matrix Θ by

$$f(x) = \sqrt{\frac{\det \Theta}{(2\pi)^p}} \exp\left\{-\frac{1}{2}x^T \Theta x\right\} \quad (x \in \mathbb{R}^p)$$

and let A, B, C be disjoint subsets of $\{1, \ldots, p\}$. Show that if

$$f_{A \cup C}(x_{A \cup C}) f_{B \cup C}(x_{B \cup C}) = f_{A \cup B \cup C}(x_{A \cup B \cup C}) f_C(x_C) \qquad \text{(cf. (5.2))}$$

for arbitrary $x \in \mathbb{R}^p$, then $\theta_{i,j} = 0$ with $i \in A$, $j \in B$ and $i \in B$, $j \in A$.

66. Suppose that $X \in \mathbb{R}^{N \times p}$ are N samples, each of which has been generated according to the p variable Gaussian distribution $N(0, \Sigma)$ ($\Sigma \in \mathbb{R}^{p \times p}$), and let $S := \frac{1}{N}X^T X$, $\Theta := \Sigma^{-1}$. Show the following statements.

(a) Suppose that $\lambda = 0$ and $N < p$. Then, no inverse matrix exists for S

$$\Theta^{-1} - S - \lambda \psi = 0 \qquad (5.30)$$

and no maximum likelihood estimate of Θ exists.

Hint The rank of S is at most N, and $S \in \mathbb{R}^{p \times p}$.

(b) The trace of $S\Theta$ with $\Theta = (\theta_{s,t})$ can be written as

$$\frac{1}{N}\sum_{i=1}^{N}\left(\sum_{s=1}^{p}\sum_{t=1}^{p}x_{i,s}\theta_{s,t}x_{i,t}\right)$$

Hint If the multiplications AB and BA of matrices A, B are defined, they have an equal trace, and

$$\text{trace}(S\Theta) = \frac{1}{N}\text{trace}(X^{T}X\Theta) = \frac{1}{N}\text{trace}(X\Theta X^{T})$$

when $A = X^{T}$ and $B = X\Theta$.

(c) If we express the probability density function of the p Gaussian variables as $f_{\Theta}(x_{1}, \ldots, x_{p})$, then the log-likelihood $\frac{1}{N}\sum_{i=1}^{N}\log f_{\Theta}(x_{i,1}, \ldots, x_{i,p})$ can be written as

$$\frac{1}{2}\{\log\det\Theta - p\log(2\pi) - \text{trace}(S\Theta)\} . \qquad\qquad \text{(cf. (5.9))}$$

(d) For $\lambda \geq 0$, the maximization of $\frac{1}{N}\sum_{i=1}^{N}\log f_{\Theta}(x_{i}) - \frac{1}{2}\lambda\sum_{s\neq t}|\theta_{s,t}|$ w.r.t. Θ is

that of

$$\log\det\Theta - \text{trace}(S\Theta) - \lambda\sum_{s\neq t}|\theta_{s,t}| \qquad \text{(cf. (5.10))} . \qquad (5.31)$$

67. Let $A_{i,j}$ and $|B|$ be the submatrix that excludes the i-th row and j-th column from matrix $A \in \mathbb{R}^{p\times p}$ and the determinant of $B \in \mathbb{R}^{m\times m}$ ($m \leq p$), respectively. Then, we have

$$\sum_{j=1}^{p}(-1)^{k+j}a_{i,j}|A_{k,j}| = \begin{cases} |A|, & i = k \\ 0, & i \neq k \end{cases} . \qquad \text{(cf. (5.11))}$$

(a) When $|A| \neq 0$, let B be the matrix whose (j, k) element is $b_{j,k} = (-1)^{k+j}|A_{k,j}|/|A|$. Show that AB is the unit matrix. Hereafter, we write this matrix B as A^{-1}.
(b) Show that if we differentiate $|A|$ by $a_{i,j}$, then it becomes $(-1)^{i+j}|A_{i,j}|$.
(c) Show that if we differentiate $\log|A|$ by $a_{i,j}$, then it becomes $(-1)^{i+j}|A_{i,j}|/|A|$, the (j, i)-th element of A^{-1}.
(d) Show that if we differentiate the trace of $S\Theta$ by the (s, t)-element $\theta_{s,t}$ of Θ, it becomes the (t, s)-th element of S.
   Hint Differentiate $\text{trace}(S\Theta) = \sum_{i=1}^{p}\sum_{j=1}^{p}s_{i,j}\theta_{j,i}$ by $\theta_{s,t}$.
(e) Show that the Θ that maximizes (5.31) is the solution of (5.12), where $\Psi = (\psi_{s,t})$ is $\psi_{s,t} = 0$ if $s = t$ and

$$\psi_{s,t} = \begin{cases} 1, & \theta_{s,t} > 0 \\ [-1, 1], & \theta_{s,t} = 0 \\ -1, & \theta_{s,t} < 0 \end{cases}$$

otherwise.

Hint If we differentiate $\log \det \Theta$ by $\theta_{s,t}$, it becomes the (t, s)-th element of Θ^{-1} from (d). However, because Θ is symmetric, $\Theta^{-1}: \Theta^T = \Theta \implies (\Theta^{-1})^T = (\Theta^T)^{-1} = \Theta^{-1}$ is symmetric as well.

68. Suppose that we have

$$\Psi = \begin{bmatrix} \Psi_{1,1} & \psi_{1,2} \\ \psi_{2,1} & \psi_{2,2} \end{bmatrix}, \quad S = \begin{bmatrix} S_{1,1} & s_{1,2} \\ s_{2,1} & s_{2,2} \end{bmatrix}, \quad \Theta = \begin{bmatrix} \Theta_{1,1} & \theta_{1,2} \\ \theta_{2,1} & \theta_{2,2} \end{bmatrix}$$

$$\begin{bmatrix} W_{1,1} & w_{1,2} \\ w_{2,1} & w_{2,2} \end{bmatrix} \begin{bmatrix} \Theta_{1,1} & \theta_{1,2} \\ \theta_{2,1} & \theta_{2,2} \end{bmatrix} = \begin{bmatrix} I_{p-1} & 0 \\ 0 & 1 \end{bmatrix} \quad \text{(cf. (5.13))} \qquad (5.32)$$

for $\Psi, \Theta, S \in \mathbb{R}^{p \times p}$ and W such that $W\Theta = I$, where the upper-left part is $(p-1) \times (p-1)$ and we assume that $\theta_{2,2} > 0$.

(a) Derive

$$w_{1,2} - s_{1,2} - \lambda\psi_{1,2} = 0 \quad \text{(cf. (5.14))} \qquad (5.33)$$

and

$$W_{1,1}\theta_{1,2} + w_{1,2}\theta_{2,2} = 0 \quad \text{(cf. (5.15))} \qquad (5.34)$$

from the upper-right part of (5.30) and (5.32), respectively.

Hint The upper-right part of Θ^{-1} is $w_{1,2}$.

(b) Let $\beta = \begin{bmatrix} \beta_1 \\ \vdots \\ \beta_{p-1} \end{bmatrix} := -\dfrac{\theta_{1,2}}{\theta_{2,2}}$. Show that from (5.33), (5.34), the two equations are obtained as

$$W_{1,1}\beta - s_{1,2} + \lambda\phi_{1,2} = 0 \quad \text{(cf. (5.16))} \qquad (5.35)$$

$$w_{1,2} = W_{1,1}\beta \quad \text{(cf. (5.17))}, \qquad (5.36)$$

where the j-th element of $\phi_{1,2} \in \mathbb{R}^{p-1}$ is $\begin{cases} 1, & \beta_j > 0 \\ [-1, 1], & \beta_j = 0 \\ -1, & \beta_j < 0 \end{cases}$.

Hint The sign of each element of $\psi_{2,2} \in \mathbb{R}^{p-1}$ is determined by the sign of $\theta_{1,2}$. From $\theta_{2,2} > 0$, the signs of β and $\theta_{1,2}$ are opposite; thus, $\psi_{1,2} = -\phi_{1,2}$.

69. We obtain the solution of (5.30) in the following way: find the solution of (5.35) w.r.t. β, substitute it into (5.36), and obtain $w_{1,2}$, which is the same as $w_{2,1}$ due to symmetry.

We repeat the process, changing the positions of $W_{1,1}$, $w_{1,2}$. For $j = 1, \ldots, p$, $W_{1,1}$ is regarded as W except in the j-th row and j-th column, and $w_{1,2}$ is regarded as W in the j-th column except for the j-th element.

In the last stage, for $j = 1, \ldots, p$, if we take the following one cycle, we obtain the estimate of Θ:

$$\theta_{2,2} = [w_{2,2} - w_{1,2}\beta]^{-1} \quad \text{(cf. (5.19))} \tag{5.37}$$

$$\theta_{1,2} = -\beta\theta_{2,2} \quad \text{(cf. (5.20))} \tag{5.38}$$

(a) Let $\quad A := W_{1,1} \in \mathbb{R}^{(p-1)\times(p-1)}, \quad b = s_{1,2} \in \mathbb{R}^{p-1}, \quad$ and $\quad c_j := b_j - \sum_{k \neq j} a_{j,k}\beta_k$. Show that each β_j that satisfies (5.35) can be computed via

$$\beta_j = \begin{cases} \dfrac{c_j - \lambda}{a_{j,j}}, & c_j > \lambda \\ 0, & -\lambda < c_j < \lambda \\ \dfrac{c_j + \lambda}{a_{j,j}}, & c_j < -\lambda \end{cases} \quad \text{(cf. (5.18))} \tag{5.39}$$

(b) Derive (5.37) from (5.32).

70. We construct the graphical Lasso as follows.

```
1  library(glasso)
2  solve(s); glasso(s, rho=0); glasso(s, rho = 0.01)
```

Execute each step, and compare the results.

```
1  inner.prod = function(x, y) return(sum(x * y))
2  soft.th = function(lambda, x) return(sign(x) * pmax(abs(x) -
       lambda, 0))
3  graph.lasso = function(s, lambda = 0) {
4    W = s; p = ncol(s); beta = matrix(0, nrow = p - 1, ncol = p)
5    beta.out = beta; eps.out = 1
6    while (eps.out > 0.01) {
7      for (j in 1:p) {
8        a = W[-j, -j]; b = s[-j, j]
9        beta.in = beta[, j]; eps.in = 1
10       while (eps.in > 0.01) {
11         for (h in 1:(p - 1)) {
12           cc = b[h] - inner.prod(a[h, -h], beta[-h, j])
13           beta[h, j] = soft.th(lambda, cc) / a[h, h]
14         }
15         eps.in = max(beta[, j] - beta.in); beta.in = beta[, j]
16       }
17       W[-j, j] = W[-j, -j] %*% beta[, j]
18     }
19     eps.out = max(beta - beta.out); beta.out = beta
20   }
21   theta = matrix(nrow = p, ncol = p)
22   for (j in 1:p) {
23     theta[j, j] = 1 / (W[j, j] - W[j, -j] %*% beta[, j])
24     theta[-j, j] = -beta[, j] * theta[j, j]
```

```
25    }
26    return(theta)
27 }
```

What rows of the function definition of graph.lasso are the Eqs. (5.36)–(5.39)?

Then, we can construct an undirected graph G by connecting each s, t such that $\theta_{s,t} \neq 0$ as an edge. Generate the data for $p = 5$ and $N = 20$ based on a matrix $\Theta = (\theta_{s,t})$ known a priori.

```
1  library(MultiRNG)
2  Theta = matrix(c(    2,   0.6,      0,      0,  0.5,   0.6,     2, -0.4,
       0.3,    0,     0,
3                    -0.4,     2, -0.2,      0,     0,   0.3, -0.2,     2,
       -0.2,   0.5,    0,
4                       0, -0.2,      2), nrow = 5)
5  Sigma = ## Blank ##; meanvec = rep(0, 5)
6  # mean: mean.vec, cov matrix: cov.mat, samples #: no.row, variable
         #: d
7  # Generate the sample matrix
8  dat = draw.d.variate.normal(no.row = 20, d = 5, mean.vec = meanvec
       , cov.mat = Sigma)
```

Moreover, execute the following code, and examine whether the precision matrix Θ is correctly estimated.

```
1  s = t(dat) %*% dat / nrow(dat); graph.lasso(s); graph.lasso(s,
       lambda = 0.01)
```

71. The function adj defined below connects each (i, j) as an edge if and only if the element is nonzero given a symmetric matrix of size p to construct an undirected graph. Execute it for the breastcancer.csv data.

```
1  library(igraph)
2  adj = function(mat) {
3    p = ncol(mat); ad = matrix(0, nrow = p, ncol = p)
4    for (i in 1:(p - 1)) for (j in (i + 1):p) {
5      if (## Blank ##) {ad[i, j] = 0} else {ad[i, j] = 1}
6    }
7    g = graph.adjacency(ad, mode = "undirected")
8    plot(g)
9  }
10
11 library(glasso)
12 df = read.csv("breastcancer.csv")
13 w = matrix(nrow = 250, ncol = 1000)
14 for (i in 1:1000) w[, i] = as.numeric(df[[i]])
15 x = w; s = t(x) %*% x / 250
16 fit = glasso(s, rho = 1); sum(fit$wi == 0); adj(fit$wi)
17 fit = glasso(s, rho = 0.5); sum(fit$wi == 0); adj(fit$wi)
18 y = NULL; z = NULL
19 for (i in 1:999) for (j in (i + 1):1000) {
20   if (mat[i, j] != 0) {y = c(y, i); z = c(z, j)}
21 }
```

```
22  cbind(y,z)
23  ## Restrict to the first 100 genes
24  x = x[, 1:100]; s = t(x) %*% x / 250
25  fit = glasso(s, rho = 0.75); sum(fit$wi == 0); adj(fit$wi)
26  y = NULL; z = NULL
27  for (i in 1:99) for (j in (i + 1):100) {
28      if (fit$wi[i, j] != 0) {y = c(y, i); z = c(z, j)}
29  }
30  cbind(y, z)
31  fit = glasso(s, rho = 0.25); sum(fit$wi == 0); adj(fit$wi)
```

72. The following code generates an undirected graph via the quasi-likelihood method and the `glmnet` package. We examine the difference between using the AND and OR rules, where the original data contains $p = 1,000$ but the execution is for the first $p = 50$ genes to save time. Execute the OR rule case as well as the AND case by modifying the latter.

```
1   library(glmnet)
2   df = read.csv("breastcancer.csv")
3   n = 250; p = 50; w = matrix(nrow = n, ncol = p)
4   for (i in 1:p) w[, i] = as.numeric(df[[i]])
5   x = w[, 1:p]; fm = rep("gaussian", p); lambda = 0.1
6   fit = list()
7   for (j in 1:p) fit[[j]] = glmnet(x[, -j], x[, j], family = fm[j],
        lambda = lambda)
8   ad = matrix(0, p, p)
9   for (i in 1:p) for (j in 1:(p - 1)) {
10      k = j
11      if (j >= i) k = j + 1
12      if (fit[[i]]$beta[j] != 0) {ad[i, k] = 1} else {ad[i, k] = 0}
13  }
14  ## AND
15  for (i in 1:(p - 1)) for (j in (i + 1):p) {
16      if (ad[i, j] != ad[i, j]) {ad[i, j] = 0; ad[j, i] = 0}
17  }
18  u = NULL; v = NULL
19  for (i in 1:(p - 1)) for (j in (i + 1):p) {
20      if (ad[i, j] == 1) {u = c(u, i); v = c(v, j)}
21  }
22  ## OR
```

We execute the quasi-likelihood method for the signs ± of breastcancer.csv.

```
1  library(glmnet)
2  df = read.csv("breastcancer.csv")
3  w = matrix(nrow = 250, ncol = 1000); for (i in 1:1000) w[, i] = as.
       numeric(df[[i]])
4  w = (sign(w) + 1) / 2  ## Transform to Binary
5  p = 50; x = w[, 1:p]; fm = rep("binomial", p); lambda = 0.15
6  fit = list()
7  for (j in 1:p) fit[[j]] = glmnet(x[, -j], x[, j], family = fm[j],
       lambda = lambda)
8  ad = matrix(0, nrow = p, ncol = p)
9  for (i in 1:p) for (j in 1:(p - 1)) {
10   k = j
11   if (j >= i) k = j + 1
12   if (fit[[i]]$beta[j] != 0) {ad[i, k] = 1} else {ad[i, k] = 0}
13 }
14 for (i in 1:(p - 1)) for (j in (i + 1):p) {
15   if (ad[i, j] != ad[i, j]) {ad[i, j] = 0; ad[j, i] = 0}
16 }
17 sum(ad); adj(ad)
```

How can we deal with data that contain both continuous and discrete values?

73. Joint graphical Lasso (JGL) finds $\Theta_1, \ldots, \Theta_K$ that maximize

$$\sum_{i=1}^{K} N_k \{\log \det \Theta_k - \text{trace}(S\Theta_k)\} - P(\Theta_1, \ldots, \Theta_K) \qquad \text{(cf. (5.21))} \quad (5.40)$$

given $X \in \mathbb{R}^{N \times p}$, $y \in \{1, 2, \ldots, K\}^N$ ($K \geq 2$). Suppose that $P(\Theta_1, \ldots, \Theta_K)$ expresses the fused Lasso penalty

$$P(\Theta_1, \ldots, \Theta_K) := \lambda_1 \sum_{k} \sum_{i \neq j} |\theta_{i,j}^{(k)}| + \lambda_2 \sum_{k < k'} \sum_{i,j} |\theta_{i,j}^{(k)} - \theta_{i,j}^{(k')}|, \quad \text{(cf. (5.22))}$$

where the indices in $\sum$ range over $k = 1, \ldots, K$, $i, j = 1, \ldots, p$. In JGL, to obtain the solution of (5.40), we apply the ADMM that was introduced in Chap. 4. We define the extended Lagrangian as

$$L_\rho(\Theta, Z, U) := -\sum_{k=1}^{K} N_k \{\log \det \Theta_k - \text{trace}(S\Theta_k)\}$$

$$+ P(Z_1, \ldots, Z_K) + \rho \sum_{k=1}^{K} < U_k, \Theta_k - Z_k > + \frac{\rho}{2} \sum_{k=1}^{K} \|\Theta_k - Z_k\|_F^2$$

$$\text{(cf. (5.23))}$$

and repeat the following steps:

i. $\Theta^{(t)} \leftarrow \text{argmin}_\Theta L_\rho(\Theta, Z^{(t-1)}, U^{(t-1)})$

ii. $Z^{(t)} \leftarrow \text{argmin}_Z L_\rho(\Theta^{(t)}, Z, U^{(t-1)})$

iii. $U^{(t)} \leftarrow U^{(t-1)} + \rho(\Theta^{(t)} - Z^{(t)})$

More precisely, setting $\rho > 0$ and letting Θ_k and Z_k, U_k be the unit and zero matrices $k = 1, \ldots, K$, respectively, repeat the above three steps.

(a) Show that if we differentiate (5.40) by Θ_k in the first step, then we have

$$-N_k(\Theta_k^{-1} - S_k) + \rho(\Theta_k - Z_k + U_k) = 0 .$$

(b) We wish to obtain the optimum Θ_k in the first step. To this end, we decompose both sides of the symmetric matrix

$$\Theta_k^{-1} - \frac{\rho}{N_k}\Theta_k = S_k - \rho\frac{Z_k}{N_k} + \rho\frac{U_k}{N_k}$$

as $V D V^T$ and obtain $\tilde{D}$ such that

$$\tilde{D}_{j,j} = \frac{N_k}{2\rho}(-D_{j,j} + \sqrt{D_{j,j}^2 + 4\rho/N_k})$$

from the diagonal matrix D. Show that $V \tilde{D} V^T$ is the optimum Θ.

(c) In the second step, for $K = 2$, we require a fused Lasso procedure for two values. Let y_1, y_2 be the two sets of data. Derive θ_1, θ_2 that minimize

$$\frac{1}{2}(y_1 - \theta_1)^2 + \frac{1}{2}(y_2 - \theta_2)^2 + |\theta_1 - \theta_2| .$$

74. We construct the fused Lasso JGL. Fill in the blanks, and execute the procedure.

```
1   ## Fused Lasso for size 2
2   b.fused = function(y, lambda) {
3       if (y[1] > y[2] + 2 * lambda) {a = y[1] - lambda; b = y[2] +
            lambda}
4       else if (y[1] < y[2] - 2 * lambda) {a = y[1] + lambda; b = y[2]
            - lambda}
5       else {a = (y[1] + y[2]) / 2; b = a}
6       return(c(a, b))
7   }
8   ## Fused Lasso that compares all the adjacency values
9   fused = function(y, lambda.1, lambda.2) {
10      K = length(y)
11      if (K == 1) {theta = y}
12      else if (K == 2) {theta = b.fused(y, lambda.2)}
13      else {
14          L = K * (K - 1) / 2; D = matrix(0, nrow = L, ncol = K)
15          k = 0
16          for (i in 1:(K - 1)) for (j in (i + 1):K) {
17              k = k + 1; ## Blank (1) ##
18          }
19          out = genlasso(y, D = D)
```

```
20        theta = coef(out, lambda = lambda.2)
21      }
22      theta = soft.th(lambda.1, theta)
23      return(theta)
24    }
25    ## Joint Graphical Lasso
26    jgl = function(X, lambda.1, lambda.2) {   # X is given as a list
27      K = length(X); p = ncol(X[[1]]); n = array(dim = K); S = list()
28      for (k in 1:K) {n[k] = nrow(X[[k]]); S[[k]] = t(X[[k]]) %*% X[[k
          ]] / n[k]}
29      rho = 1; lambda.1 = lambda.1 / rho; lambda.2 = lambda.2 / rho
30      Theta = list(); for (k in 1:K) Theta[[k]] = diag(p)
31      Theta.old = list(); for (k in 1:K) Theta.old[[k]] = diag(rnorm(p
          ))
32      U = list(); for (k in 1:K) U[[k]] = matrix(0, nrow = p, ncol = p
          )
33      Z = list(); for (k in 1:K) Z[[k]] = matrix(0, nrow = p, ncol = p
          )
34      epsilon = 0; epsilon.old = 1
35      while (abs((epsilon - epsilon.old) / epsilon.old) > 0.0001) {
36        Theta.old = Theta; epsilon.old = epsilon
37        ## Update (i)
38        for (k in 1:K) {
39          mat = S[[k]] - rho * Z[[k]] / n[k] + rho * U[[k]] / n[k]
40          svd.mat = svd(mat)
41          V = svd.mat$v
42          D = svd.mat$d
43          DD = ## Blank (2) ##
44          Theta[[k]] = ## Blank (3) ##
45        }
46        ## Update (ii)
47        for (i in 1:p) for (j in 1:p) {
48          A = NULL; for (k in 1:K) A = c(A, Theta[[k]][i, j] + U[[k]][
            i, j])
49          if (i == j) {B = fused(A, 0, lambda.2)} else {B = fused(A,
            lambda.1, lambda.2)}
50          for (k in 1:K) Z[[k]][i,j] = B[k]
51        }
52        ## Update (iii)
53        for (k in 1:K) U[[k]] = U[[k]] + Theta[[k]] - Z[[k]]
54        ## test convergence
55        epsilon = 0
56        for (k in 1:K) {
57          epsilon.new = max(abs(Theta[[k]] - Theta.old[[k]]))
58          if (epsilon.new > epsilon) epsilon = epsilon.new
59        }
60      }
61      return(Z)
62    }
63
64    ## Data Generation and Execute
65    p = 10; K = 2; N = 100; n = array(dim = K); for (k in 1:K) n[k] =
          N / K
66    X = list(); X[[1]] = matrix(rnorm(n[k] * p), ncol = p)
67    for (k in 2:K) X[[k]] = X[[k - 1]] + matrix(rnorm(n[k] * p) * 0.1,
          ncol = p)
```

```
68   ## Change lambda.1, lambda.2 and Execute
69   Theta = jgl(X, 3, 0.01)
70   par(mfrow = c(1, 2)); adj(Theta[[1]]); adj(Theta[[2]])
```

75. For the group Lasso, only the second step should be replaced. Let $A_k[i, j] = \Theta_k[i, j] + U_k[i, j]$. Then, no update is required for $i = j$, and

$$
Z_k[i, j] = S_{\lambda_1/\rho}(A_k[i, j]) \left(1 - \frac{\lambda_2}{\rho \sqrt{\sum_{k=1}^{K} S_{\lambda_1/\rho}(A_k[i, j])^2}} \right)_+
$$

for $i \neq j$. We construct the code. Fill in the blank, and execute the procedure as in the previous exercise.

```
1   for (i in 1:p) for (j in 1:p) {
2       A = NULL; for (k in 1:K) A = c(A, Theta[[k]][i, j] + U[[k]][i, j
        ])
3       if (i == j) {B = A} else {B = ## Blank ##}
4       for (k in 1:K) Z[[k]][i, j] = B[k]
5   }
```

Chapter 6
Matrix Decomposition

Thus far, by Lasso, we mean making the estimated coefficients of some variables zero if the absolute values are significantly small for regression, classification, and graphical models. In this chapter, we consider dealing with matrices. Suppose that the given data take the form of a matrix, such as in image processing. We wish to approximate an image by a low-rank matrix after singular decomposition. However, the function needed to obtain the rank from a matrix is not convex. In this chapter, we introduce matrix norms w.r.t. the singular values, such as the nuclear norm and spectral norm, and formulate the problem as convex optimization, as we did for other Lasso classes.

We list several formulas of matrix arithmetic, where we write the unit matrix of size $n \times n$ as I_n.

i. By $\|Z\|_F$, we denote the Frobenius norm of matrix Z, the square root of the square sum of the elements in $Z \in \mathbb{R}^{m \times n}$. In general, we have

$$\|Z\|_F^2 = \text{trace}(Z^T Z) . \tag{6.1}$$

In fact, because the (i, j)-th element $w_{i,j}$ of $Z^T Z$ is $\sum_{k=1}^n z_{k,i} z_{k,j}$, we have

$$\text{trace}(Z^T Z) = \sum_{i=1}^n w_{i,i} = \sum_{i=1}^n \sum_{k=1}^m z_{k,i}^2 = \|Z\|_F^2 .$$

ii. For $l, m, n \geq 1$ and matrices $A = (a_{i,j}) \in \mathbb{R}^{l \times m}$, $B = (b_{i,j}) \in \mathbb{R}^{m \times n}$, $C = (c_{i,j}) \in \mathbb{R}^{n \times l}$, the (i, j)-th elements of ABC, BCA are $u_{i,j} := \sum_{k=1}^m \sum_{h=1}^n a_{i,k} b_{k,h} c_{h,j}$ and $v_{i,j} := \sum_{k=1}^n \sum_{h=1}^l b_{i,k} c_{k,h} a_{h,j}$. If we take their traces, we have

$$\text{trace}(ABC) = \sum_{i=1}^{l} u_{i,i} = \sum_{i=1}^{l}\sum_{k=1}^{m}\sum_{h=1}^{n} a_{i,k}b_{k,h}c_{h,i} = \sum_{i=1}^{m} v_{i,i} = \text{trace}(BCA) \ .$$

(6.2)

iii. We can choose mutually orthogonal vectors as the basis of a vector space. If the lengths of the vectors in the mutually orthogonal basis are one, then the basis is said to be orthonormal. The columns $u_1, \cdots, u_n \in \mathbb{R}^n$ of $U \in \mathbb{R}^{n \times n}$ making an orthonormal basis of the vector space $\mathbb{R}^n$ are equivalent to $U^T U = U U^T = I_n$. Moreover, if $U = [u_1, \cdots, u_n]$ consists of orthonormal column vectors, for $U_r = [u_{i(1)}, \cdots, u_{i(r)}] \in \mathbb{R}^{n \times r}$ with $r \le n$ and $1 \le i(1) \le \cdots \le i(r) \le n$, $U_r^T U_r \in \mathbb{R}^{r \times r}$ is the unit matrix, but $U_r U_r^T \in \mathbb{R}^{n \times n}$ is not.

6.1 Singular Decomposition

Let $m \ge n$ and $Z \in \mathbb{R}^{m \times n}$. If we write the eigenvalue and eigenvector of nonnegative definite matrix $Z^T Z \in \mathbb{R}^{n \times n}$ as $\lambda_1 \ge \cdots \ge \lambda_n \in \mathbb{R}$ and $v_1, \ldots, v_n \in \mathbb{R}^n$, respectively, we have

$$Z^T Z v_i = \lambda_i v_i \ ,$$

where $v_1, \ldots, v_n \in \mathbb{R}^n$ have the unit length. We assume that if some eigenvalues coincide, we orthogonalize them, which means that $v_1, \ldots, v_n \in \mathbb{R}^n$ comprise an orthonormal basis. Then, we have

$$Z^T Z V = V D^2 \ ,$$

where $V = [v_1, \ldots, v_n] \in \mathbb{R}^{n \times n}$ and D is a diagonal matrix with the elements $d_1 := \sqrt{\lambda_1}, \ldots, d_n := \sqrt{\lambda_n}$. From $V V^T = I_n$, we have

$$Z^T Z = V D^2 V^T \ .$$

(6.3)

Suppose $d_n > 0$, which means that D is the inverse matrix. We define $U \in \mathbb{R}^{m \times n}$ as

$$U := Z V D^{-1} \ .$$

Then, Z can be written as $Z = U D V^T$, which we refer to as the singular decomposition of Z. From (6.3) and $V^T V = I_n$, we have

$$U^T U = (Z V D^{-1})^T Z V D^{-1} = D^{-1} V^T Z^T Z V D^{-1} = D^{-1} V^T (V D^2 V^T) V D^{-1} = I_n \ .$$

The ranks of $Z^T Z$ and Z coincide. In fact, for arbitrary $x \in \mathbb{R}^n$, we have

$$Zx = 0 \implies Z^T Z x = 0$$

$$Z^T Z x = 0 \implies x^T Z^T Z x = 0 \implies \|Zx\|^2 = 0 \implies Zx = 0$$

the kernels of $Z \in \mathbb{R}^{m \times n}$, $Z^T Z \in \mathbb{R}^{n \times n}$ coincide, and the numbers of columns are equal. Thus, the r value such that $d_r > 0$, $d_{r+1} = 0$ coincides with the rank of Z.

We now consider a general case $r \le n$. Let $V_r := [v_1, \ldots, v_r] \in \mathbb{R}^{n \times r}$ and $D_r \in \mathbb{R}^{r \times r}$ be a diagonal matrix with the elements $d_1, \ldots, d_r$. Then, if we regard the row vectors $d_1 v_1^T, \ldots, d_r v_r^T \in \mathbb{R}^n$ of $D_r V_r^T \in \mathbb{R}^{r \times n}$ (rank: r) as a basis of the rows $z_i \in \mathbb{R}^n$ ($i = 1, \ldots, m$) of Z, we find that $u_{i,j} \in \mathbb{R}$ exists and is unique such that $z_i = \sum_{j=1}^{r} u_{i,j} d_j v_j^T$. Thus, $U_r = (u_{i,j}) \in \mathbb{R}^{m \times r}$ such that $Z = U_r D_r V_r^T$ is unique.

Moreover, (6.3) means that $U_r^T U_r = I_r$. In fact, let V_{n-r} be the submatrix of V that excludes V_r. From $V_r^T V_r = I_r$ and $V_r^T V_{n-r} = O \in \mathbb{R}^{r \times (n-r)}$, we have

$$Z^T Z = V D^2 V^T \iff V_r D_r U_r^T U_r D_r V_r^T = V D^2 V^T$$
$$\iff V^T V_r D_r U_r^T U_r D_r V_r^T V = D^2$$
$$\iff \begin{bmatrix} D_r \\ O \end{bmatrix} U_r^T U_r [D_r \quad O] = \begin{bmatrix} D_r^2 & O \\ O & O \end{bmatrix}$$
$$\iff U_r^T U_r = I_r .$$

Thus, the decomposition is unique up to the choice of the basis of the eigenspace V (we often write this as $Z = \sum_{i=1}^{r} d_i u_i v_i^T$).

If $Z \in \mathbb{R}^{m \times n}$ ($m < n$), after obtaining $\bar{U} \in \mathbb{R}^{n \times m}$, $D \in \mathbb{R}^{m \times m}$, and $\bar{V} \in \mathbb{R}^{m \times m}$ such that $Z^T = \bar{U} D \bar{V}^T$, we rewrite it as $Z = (\bar{U} D \bar{V}^T)^T = \bar{V} D \bar{U}^T$. If we write Z as $Z = U D V^T$, then $U = \bar{V} \in \mathbb{R}^{m \times m}$, $D \in \mathbb{R}^{m \times m}$, $V = \bar{U} \in \mathbb{R}^{n \times m}$ such that V is not square.

Example 53 For $Z = \begin{bmatrix} 0 & -2 \\ 5 & -4 \\ -1 & 1 \end{bmatrix}$, we execute the singular decompositions of Z and Z^T using the following code. If we transpose Z, then U and V are switched.

```
z = matrix(c(0, 5, -1, -2, -4, 1), nrow = 3); z
```

```
     [,1] [,2]
[1,]    0   -2
[2,]    5   -4
[3,]   -1    1
```

```
svd(z)
```

```
1   $d
2   [1] 6.681937 1.533530
3   $u
4              [,1]          [,2]
5   [1,] -0.1987442  0.97518046
6   [2,] -0.9570074 -0.21457290
7   [3,]  0.2112759 -0.05460347
8   $v
9              [,1]          [,2]
10  [1,] -0.7477342 -0.6639982
11  [2,]  0.6639982 -0.7477342
```

```
1   svd(t(Z))
```

```
1   $d
2   [1] 6.681937 1.533530
3   $u
4              [,1]         [,2]
5   [1,] -0.7477342 0.6639982
6   [2,]  0.6639982 0.7477342
7   $v
8              [,1]          [,2]
9   [1,] -0.1987442 -0.97518046
10  [2,] -0.9570074  0.21457290
11  [3,]  0.2112759  0.05460347
```

Example 54 If $Z \in \mathbb{R}^{n \times n}$ is symmetric, the eigenvalues of $Z^T Z = Z^2$ are the squares of the eigenvalues of Z. In fact, from

$$\det(Z^2 - tI) = \det(Z - \sqrt{t}I)\det(Z + \sqrt{t}I) ,$$

the solutions $\lambda_1, \ldots, \lambda_n$ for $\pm\sqrt{t}$ of $\det(Z - \sqrt{t}I) = 0$, $\det(Z + \sqrt{t}I) = 0$ are the eigenvalues of Z, and the solution for $t \geq 0$ of $\det(Z^2 - tI) = 0$ is their square. For $t < 0$, $Z^2 - tI$ is positive definite, and $\det(Z^2 - tI) > 0$.

On the other hand, if we singular-decompose Z, then D is a diagonal matrix with the elements $|\lambda_1|, \ldots, |\lambda_n|$ (the absolute values of the eigenvalues of Z). Thus, if we compare the eigenvalue decomposition $Z = WDW^T$ and singular-value decomposition $Z = UDV^T$, then we have

$$\lambda_i \geq 0 \iff u_i = v_i = w_i$$
$$\lambda_i < 0 \iff u_i = w_i, \ v_i = -w_i \text{ or } u_i = -w_i, \ v_i = w_i,$$

where $w_i \in \mathbb{R}^n$ is the eigenvector associated with u_i, v_i. If Z is nonnegative definite, these decompositions coincide. In the following, we execute singular-value and eigenvalue decomposition of the matrix $Z = \begin{bmatrix} 0 & 5 \\ 5 & -1 \end{bmatrix}$.

```
z = matrix(c(0, 5, 5, -1), nrow = 2); z
```

```
     [,1] [,2]
[1,]    0    5
[2,]    5   -1
```

```
svd(z)
```

```
$d
[1] 5.524938 4.524938
$u
              [,1]        [,2]
[1,]    0.6710053 -0.7414525
[2,]   -0.7414525 -0.6710053
$v
              [,1]        [,2]
[1,]   -0.6710053 -0.7414525
[2,]    0.7414525 -0.6710053
```

```
eigen(z)
```

```
$values
[1]   4.524938 -5.524938
$vectors
              [,1]        [,2]
[1,]   -0.7414525 -0.6710053
[2,]   -0.6710053  0.7414525
```

6.2 Eckart-Young's Theorem

We herein state Eckart-Young's theorem, which plays a crucial role in approximating a low-rank alternative matrix.

Proposition 18 (Eckart-Young's Theorem) *Let $m \leq n$. We are given observation $Z = UDV^T \in \mathbb{R}^{m \times n}$, where $U \in \mathbb{R}^{m \times m}$, $V \in \mathbb{R}^{m \times n}$, and the rank of Z is m. The $M \in \mathbb{R}^{m \times n}$ with the rank of at most r that minimizes $\|Z - M\|_F$ is given by UD_rV^T, where $D_r \in \mathbb{R}^{m \times m}$ is obtained by making the elements $d_{r+1}, \cdots, d_n$ in $D \in \mathbb{R}^{m \times m}$ zero.*

For the proof, see the Appendix.

We can construct a function svd.r that approximates a matrix z by rank r as follows. It diminishes only the diagonal elements of D, except the first d elements.

Fig. 6.1 We compared the matrix obtained after executing svd.r(z,r) and the original z. The horizontal and vertical axes show *r* and the difference in the squared Frobenius norms

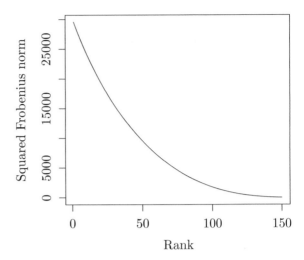

```
1   svd.r = function(z, r) {
2     n = min(nrow(z), ncol(z))
3     ss = svd(z)
4     tt = ss$u %*% diag(c(ss$d[1:r], rep(0, n - r))) %*% t(ss$v)
5     return(tt)
6   }
```

Example 55 Executing the function svd.r, we examine the behavior. We show the output in Fig. 6.1.

```
1   m = 200; n = 150; z = matrix(rnorm(m * n), nrow = m)
2   F.norm = NULL; for (r in 1:n) {m = svd.r(z, r); F.norm = c(F.norm, norm
       (z - m, "F") ^ 2)}
3   plot(1:n, F.norm, type = "l", xlab = "Rank", ylab = "Squared Frobenius
       Norm")
```

Example 56 The following code is used to obtain another image file with a lower rank from the original lion.jpg. In general, if an image takes the *JPG* form, then each pixel takes 256 grayscale values. Moreover, if the image is in color, we have the information for each of the blue, red, and green pixels. For example, if the display is 480×360 pixels in size, then it has $480 \times 360 \times 3$ pixels in total. The following codes are executed for ranks $r = 2, 5, 10, 20, 50, 100$ (Fig. 6.2). Before executing the procedure, we need to generate a folder /compressed under the current folder beforehand.

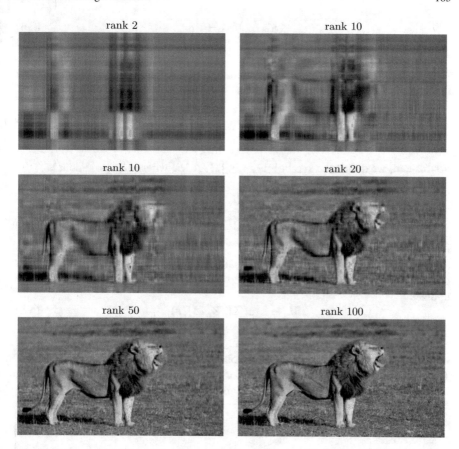

Fig. 6.2 We approximate the red, green, and blue matrix expression of the image file lion.jpg by several ranks r. We observe that the lower the rank is, the less clear the approximated image

```
1  library(jpeg)
2  image = readJPEG('lion.jpg')
3  rank.seq = c(2, 5, 10, 20, 50, 100)
4  mat = array(0, dim = c(nrow(image), ncol(image), 3))
5  for (j in rank.seq) {
6    for (i in 1:3) mat[, , i] = svd.r(image[, , i], j)
7    writeJPEG(mat, paste("compressed/lion_compressed", "_mat_rank_", j, "
       .jpg", sep = ""))
8  }
```

For the next section, we assume the following situation. Suppose that the locations indexed by $\Omega \subseteq \{1, \ldots, m\} \times \{1, \ldots, n\}$ in the matrix $Z \in \mathbb{R}^{m \times n}$ are observed. Given $z_{i,j}$ $((i, j) \in \Omega)$, we wish to contain the matrix $M \in \mathbb{R}^{m \times n}$ with a rank of at most r that minimizes the Frobenius norm of $M - Z$, i.e., the M that minimizes

$$\|Z - M\|_F^2 + \lambda r \qquad (6.4)$$

for some $\lambda > 0$.

If we fix the rank r (which is equivalent to fixing some λ), the problem is to find the matrix M with rank r that minimizes

$$\sum_{(i,j)\in\Omega} (z_{i,j} - m_{i,j})^2$$

To this end, we define the symbol $[\Omega, \cdot, \cdot]: [\Omega, A, B]$ whose (i, j)-element is

$$\begin{cases} a_{i,j}, & (i, j) \in \Omega \\ b_{i,j}, & (i, j) \notin \Omega \end{cases}$$

for matrices $A = (a_{i,j})$, $B = (b_{i,j}) \in \mathbb{R}^{m\times n}$.

We write the matrix obtained by approximating a matrix A by rank r via singular decomposition as $\mathrm{svd}(A, r)$. At first, we randomly give the matrix and repeat the updates

$$M \leftarrow \mathrm{svd}([\Omega, Z, M], r) . \qquad (6.5)$$

For example, we construct the following procedure (this process stops at a local optimum and cannot converge to the globally optimum M that minimizes (6.4)). The function mask takes one and zero for the observed and unobserved positions, respectively; thus, 1 - mask takes the opposite values.

```
1   mat.r = function(z, mask, r) {
2      z = as.matrix(z)
3      min = Inf
4      m = nrow(z); n = ncol(z)
5      for (j in 1:5) {
6         guess = matrix(rnorm(m * n), nrow = m)
7         for (i in 1:10) guess = svd.r(mask * z + (1 - mask) * guess, r)
8         value = norm(mask * (z - guess), "F")
9         if (value < min) {min.mat = guess; min = value}
10     }
11     return(min.mat)
12  }
```

Example 57 To obtain the M that minimizes (6.4), we repeat the updates (6.5). In the inner loop, the rank is approximated to r, and the unobserved value is predicted ten times. In the outer loop, to prevent falling into a local solution, the initial value is changed and executed five times to find the solution that minimizes the Frobenius norm of the difference.

```
1  library(jpeg)
2  image = readJPEG('lion.jpg')
3  m = nrow(image); n = ncol(image)
4  mask = matrix(rbinom(m * n, 1, 0.5), nrow = m)
5  rank.seq = c(2, 5, 10, 20, 50, 100)
6  mat = array(0, dim = c(nrow(image), ncol(image), 3))
7  for (j in rank.seq) {
8    for (i in 1:3) mat[, , i] = mat.r(image[, , i], mask, j)
9    writeJPEG(mat, paste("compressed/lion_compressed", "_mat_rank_", j, "
       .jpg", sep = ""))
10 }
```

The main reason for the hardness of optimization as in (6.4) is that the arithmetic of computing the rank is not convex. We show that the proposition

For arbitrary $A, B \in \mathbb{R}^{m \times n}$ and $0 < \alpha < 1$,

$$\operatorname{rank}(\alpha A + (1 - \alpha)B) \leq \alpha \cdot \operatorname{rank} A + (1 - \alpha) \operatorname{rank} B \tag{6.6}$$

is false, which means that the optimization of (6.4) is not convex.

Example 58 The following is a counterexample of (6.6). The ranks in (6.6) are two and one on the left and right sides if

$$\alpha = 0.5, \quad A = \begin{bmatrix} 1 & 0 \\ 0 & 0 \end{bmatrix}, \quad \text{and } B = \begin{bmatrix} 0 & 0 \\ 0 & 1 \end{bmatrix}.$$

6.3 Norm

We consider the function $\| \cdot \| : \mathbb{R}^n \to \mathbb{R}$ to be a norm if

$$\|\alpha a\| = |\alpha| \, \|a\|$$
$$\|a\| = 0 \iff a = 0$$
$$\|a + b\| \leq \|a\| + \|b\|$$

for $\alpha \in \mathbb{R}$, $a, b \in \mathbb{R}^n$. The L1 and L2 norms that we have considered in this book satisfy them. We note that the norm is a convex function. In fact, from the first and last conditions, for $0 \leq \alpha \leq 1$, we have

$$\|\alpha a + (1 - \alpha)b\| \leq \|\alpha a\| + \|(1 - \alpha)b\| = \alpha \|a\| + (1 - \alpha)\|b\| .$$

Then, $\|b\|_* := \sup_{\|a\| \leq 1} \langle a, b \rangle$ satisfies the above three conditions as well. In fact, if an element in b is nonzero, then we have $\langle a, b \rangle > 0$ for any a that shares its sign with that of b for each nonzero element of b. Thus, $\|b\|_* = 0 \iff b = 0$. Moreover, we have

$$\sup_{\|a\|\leq 1} \langle a, b + c \rangle \leq \sup_{\|a\|\leq 1} \langle a, b \rangle + \sup_{\|a\|\leq 1} \langle a, c \rangle \,.$$

We regard such a norm $\|\cdot\|_*$ as the dual norm of $\|\cdot\|$.

Additionally, for matrices, we can similarly define the norm and its dual form. In the following, by $d_i(M)$, we denote the i-th singular value of a matrix $M \in \mathbb{R}^{m \times n}$. Let $m \geq n$. We consider

$$\|M\|^* := \sup_{\|x\|_2 \leq 1} \|Mx\|_2$$

to be the spectral norm of a matrix M, which is also a norm:

$$\sup_{\|x\|_2 \leq 1} \|(M_1 + M_2)x\|_2 \leq \sup_{\|x\|_2 \leq 1} \|M_1 x\|_2 + \sup_{\|x\|_2 \leq 1} \|M_2 x\|_2$$

Proposition 19 *The spectral norm $\|M\|^*$ of a matrix M coincides with the largest singular value $d_1(M)$ of M.*

Proof We have

$$\sup_{\|x\|_2 \leq 1} \|Mx\|_2 = \sup_{\|x\|_2 \leq 1} \|UDV^T x\|_2 = \sup_{\|x\|_2 \leq 1} \|DV^T x\|_2 = \sup_{\|y\|_2 \leq 1} \|Dy\|_2 = d_1(M)\,,$$

where we used $\|UDV^T x\|^2 = x^T V D U^T U D V^T x = \|DV^T x\|^2$ and $y^T y = x^T V V^T x = x^T x$ for $y = V^T x$ ($y = [1, 0, \ldots, 0]^T$ satisfies the last equality). $\square$

We define the dual of the norm $\|\cdot\|^*$ for a matrix $M \in \mathbb{R}^{m \times n}$ as

$$\|M\|_* := \sup_{\|Q\|^* \leq 1} \langle M, Q \rangle \,,$$

where the inner product $\langle M, Q \rangle$ is the trace of $M^T Q$. We consider the dual $\|M\|_*$ of the spectral norm of M to be the nuclear norm.

We state an inequality that plays a significant role in deriving the conclusion of this section.

Proposition 20 *For matrices $Q, M \in \mathbb{R}^{m \times n}$, if $d_1(M), \ldots, d_n(M)$ are the singular values of M, then we have*

$$\langle Q, M \rangle \leq \|Q\|^* \sum_{i=1}^{n} d_i(M) \,,$$

where the equality holds if and only if

$$d_1(Q) = \cdots = d_r(Q) \geq d_{r+1}(Q) \geq \cdots \geq d_n(Q)$$

*and either $u_i(Q) = u_i(M)$, $v_i(Q) = v_i(M)$ or $u_i(Q) = -u_i(M)$, $v_i(Q) = -v_i(M)$
for each $i = 1, \ldots, r$, where we performed the singular decomposition of M, Q as
$M = \sum_{i=1}^{n} d_i(M)u_i(M)v_i(M)$, $Q = \sum_{i=1}^{n} d_i(Q)u_i(Q)v_i(Q)$.*

For the proof, see the Appendix.

Proposition 21 *The nuclear norm $\|M\|_*$ of matrix M coincides with the sum
$\sum_{i=1}^{n} d_i(M)$ of the singular values of M.*

Proof If the singular decomposition of M is $M = UDV^T$, from (6.2), for $Q = UV^T$, we have

$$\langle Q, M \rangle = \langle UV^T, UDV^T \rangle = \text{trace}(VU^TUDV^T) = \text{trace}(V^TVU^TUD)$$

$$= \text{trace}\, D = \sum_{i=1}^{r} d_i(M),$$

where $d_i(M)$ is the i-th singular value of M. Since the maximum eigenvalue of Q
is one, it is sufficient to show that $\sup_{d_1(Q)\leq 1}\langle Q, M \rangle \sum_{i=1}^{n} d_i(M)$. However, from
Proposition 20, we have

$$\sup_{d_1(Q)\leq 1} \langle Q, M \rangle \leq \sup_{d_1(Q)\leq 1} \sum_{i=1}^{n} d_i(M)d_1(Q) = \sum_{i=1}^{n} d_i(M),$$

which completes the proof. □

In particular, the nuclear norm $\| \cdot \|_*$ is convex. Note that Proposition 20 can be
written as $\|M\|_*\|Q\|^* \geq \langle M, Q \rangle$.

Next, we consider that the set of $G \in \mathbb{R}^{m \times n}$ such that

$$\|B\|_* \geq \|A\|_* + \langle G, B - A \rangle \tag{6.7}$$

for arbitrary $B \in \mathbb{R}^{m \times n}$ is a subderivative of $\| \cdot \|_*$ at $A \in \mathbb{R}^{m \times n}$.

Proposition 22 *Let $r \geq 1$ be the rand of A. The necessary and sufficient condition
for $G = \sum_{i=1}^{n} d_i(G)u_i(G)v_i(G)^T$ to be the subderivative of the nuclear norm $\| \cdot \|_*$
at matrix $A \in \mathbb{R}^{m \times n}$ is as follows:*

$$d_1(G) = \cdots = d_r(G) = 1$$
$$1 \geq d_{r+1}(G) \geq \cdots \geq d_n(G) \geq 0$$
$$\text{either } u_i(G) = u_i(A), v_i(G) = v_i(A) \text{ or}$$
$$u_i(G) = -u_i(A), v_i(G) = -v_i(A) \text{ for each } i = 1, \ldots, r.$$

Square We show the necessity first. Since (6.7) needs to hold even when B is the
zero matrix, we have

$$\langle G, A \rangle \geq \|A\|_* .$$

Combining this with Propositions 20 and 21, we have

$$\|A\|_* \|G\|^* \geq \|A\|_* , \tag{6.8}$$

which means that $\|G\|^* \geq 1$. On the other hand, if we substitute $B = A + u_* v_*^T$ into (6.7), where u_*, v_* are the largest-singular-value left and right vectors of G, then we have

$$\|A + u_* v_*^T\|_* \geq \|A\|_* + \langle G, u_* v_*^T \rangle .$$

Because $\| \cdot \|_*$ is convex (triangle inequality), we have

$$\|A\|_* + \|u_* v_*^T\|_* \geq \|A\|_* + \|G\|^* ,$$

which means that $\|G\|^* \leq 1$. From the equality of (6.8) and Propositions 20 and $d_1(G) = \|G\|^*$, the proof of necessity is completed.

We show the sufficiency next. If we substitute $M = B$ and $Q = G$ into Proposition 20, when $\|G\|^* = 1$, we have

$$\|B\|_* \geq \langle G, B \rangle . \tag{6.9}$$

In addition, under the derived necessary condition, we have

$$\langle G, A \rangle = \text{trace} \left(\sum_{i=1}^{r} v_i(A) d_i u_i(A)^T \sum_{j=1}^{n} u_j(G) v_j(G)^T \right) = \sum_{j=1}^{r} d_i = \|A\|_* . \tag{6.10}$$

Hence, (6.9), (6.10) means (6.7), which completes the proof. □

6.4 Sparse Estimation for Low-Rank Estimations

In the previous sections, we showed that the nuclear norm $\|M\|_* = \sum_{i=1}^{n} d_i(M)$ of M is convex and derived its subderivative. In the following, we obtain the matrix $M \in \mathbb{R}^{m \times n}$ that minimizes

$$L := \frac{1}{2} \sum_{(i,j) \in \Omega} (z_{i,j} - m_{i,j})^2 + \lambda \|M\|_* \tag{6.11}$$

for $Z \in \mathbb{R}^{m \times n}$, $\Omega \subseteq \{(i, j) \mid i = 1, \ldots, m, \ j = 1, \ldots, n\}$, and $\lambda > 0$ [19].

From Proposition 22, the subderivative G of $\| \cdot \|_*$ at $M := \sum_{i=1}^{n} d_i u_i v_i^T$ is

$$G = \sum_{i=1}^{r} u_i v_i^T + \sum_{i=r+1}^{n} \tilde{d}_i \tilde{u}_i \tilde{v}_i^T$$

when the rank of M is r, where $\tilde{u}_{r+1}, \ldots, \tilde{u}_n$ and $\tilde{v}_{r+1}, \ldots, \tilde{v}_n$ are the singular vectors associated with the singular values $1 \geq \tilde{d}_{r+1}, \cdots, \tilde{d}_n$ of G.

Thus, if we differentiate both sides of (6.11) by the matrix $M = (m_{i,j})$, we have

$$[\Omega, \; M - Z, \; 0] + \lambda G \, ,$$

where the values differentiated by $m_{i,j}$ are $\dfrac{\partial L}{\partial m_{i,j}}$.

Next, we show that when the rank of M is $r < n$, m, $\|M\|_*$ cannot be differentiated by M without subderivative. If we write $M = \sum_{i=1}^{r} u_i d_i v_i^T$, for $\epsilon > 0$, we have

$$\frac{\|M + \epsilon u_j v_j^T\|_* - \|M\|_*}{\epsilon} = \frac{\sum_{i=1}^{r} d_i + \epsilon \|u_j v_j^T\|_* - \sum_{i=1}^{r} d_i}{\epsilon} = 1$$

$$\frac{\|M - \epsilon u_j v_j^T\|_* - \|M\|_*}{-\epsilon} = \frac{\sum_{i=1}^{r} d_i - \epsilon \|u_j v_j^T\|_* - \sum_{i=1}^{r} d_i}{-\epsilon} = 1$$

for $j = 1, \ldots, r$. On the other hand, because $d_j = 0$ for $j = r+1, \ldots, n$, if we subtract $\epsilon u_j v_j^T$ from M, then because the singular values cannot be negative, one of the singular vectors is multiplied by -1, and we have $d_j = \epsilon$. If we add $\epsilon u_j v_j^T$ to M, we have $d_j = \epsilon$.

Suppose that we have observations $Z \in \mathbb{R}^{m \times n}$ only for the positions $\Omega \subseteq \{1, \ldots, m\} \times \{1, \ldots, n\}$ and that we take the subderivative of (6.11) w.r.t. $M = (m_{i,j})$ to obtain the following equation:

$$[\Omega, \; M - Z, 0] + \lambda \left(\sum_{i=1}^{r} u_i v_i^T + \sum_{k=r+1}^{n} d_k \tilde{u}_k \tilde{v}_k^T \right) = 0 \, . \tag{6.12}$$

Mazumder et al. (2010) [19] showed that when starting with an arbitrary initial value M_0 and updating M_t via

$$M_{t+1} \leftarrow S_\lambda([\Omega, Z, M_t]) \tag{6.13}$$

until convergence, both sides of M in (6.13) coincide. This matrix M satisfies (6.12), where $S_\lambda(W)$ denotes the matrix whose singular values $d_i, i = 1, \ldots, n$, are replaced with $d_i - \lambda$ if $d_i \geq \lambda$ and zeros otherwise. Hence, singular-decompose $[\Omega, Z, M]$ and subtract $\min\{\lambda, d_k\}$ from the singular values d_k so that the obtained singular values are nonnegative.

Note that (6.12) can be obtained if we set $M_{t+1} = M_t = M$ in the update (6.13). For example, we can construct the function `soft.svd` that returns $Z' = U D' V^T$ given $Z = U D V^T$, where D and D' are diagonal matrices with the diagonal elements d_k and $d'_k := \max\{d_k - \lambda, 0\}$, $k = 1, \ldots .n$.

```
1   soft.svd = function(lambda, z) {
2       n = ncol(z); ss = svd(z); dd = pmax(ss$d - lambda, 0)
3       return(ss$u %*% diag(dd) %*% t(ss$v))
4   }
```

Then, the update (6.13) can be constructed as follows.

```
1   mat.lasso = function(lambda, z, mask) {
2       z = as.matrix(z); m = nrow(z); n = ncol(z)
3       guess = matrix(rnorm(m * n), nrow = m)
4       for (i in 1:20) guess = soft.svd(lambda, mask * z + (1 - mask) *
            guess)
5       return(guess)
6   }
```

Example 59 For the same image as in Example 55, we construct a procedure such that we recover the whole image when we mask some part. More precisely, we apply the functions `soft.svd` and `mat.lasso` for $p = 0.5, 0.25$ and $\lambda = 0.5, 0.1$ to recover the image (Fig. 6.3), where the larger the value of λ is, the lower the rank.

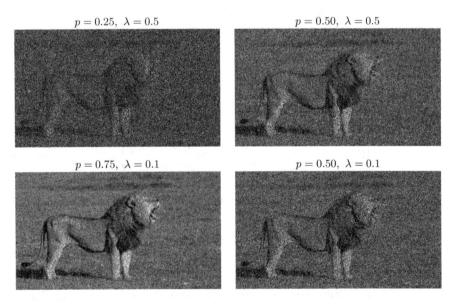

Fig. 6.3 We obtain the image that minimizes (6.11) from an image with the locations Ω of lion.jpg unmasked, where $p = 0.25, 0.5, 0.75$ are the ratios of the (unmasked) observations, λ is the parameter in (6.11), and the larger the value of λ is, the lower the rank

```
1  library(jpeg)
2  image = readJPEG('lion.jpg')
3  m = nrow(image[, , 1]); n = ncol(image[, , 1])
4  p = 0.5
5  lambda = 0.5
6  mat = array(0, dim = c(m, n, 3))
7  mask = matrix(rbinom(m * n, 1, p), ncol = n)
8  for (i in 1:3) mat[, , i] = mat.lasso(lambda, image[, , i], mask)
9  writeJPEG(mat, paste("compressed/lion_compressed", "_mat_soft.jpg", sep
       = ""))
```

Appendix: Proof of Propositions

Proposition 18 (Eckart-Young's Theorem) *Let $m \leq n$. We are given the observation $Z = UDV^T \in \mathbb{R}^{m \times n}$, where $U \in \mathbb{R}^{m \times m}$, $V \in \mathbb{R}^{m \times n}$, and the rank of Z is m. The $M \in \mathbb{R}^{m \times n}$ with a rank of at most r that minimizes $\|Z - M\|_F$ is given by UD_rV^T, where $D_r \in \mathbb{R}^{m \times m}$ is obtained by making the elements $d_{r+1}, \cdots, d_n$ in $D \in \mathbb{R}^{m \times m}$ zero.*

Proof In the following, we assume that the singular decomposition of $Z \in \mathbb{R}^{m \times n}$ ($m \leq n$) can be written as $Z = UDV^T$, where ($U \in \mathbb{R}^{m \times m}$, $D \in \mathbb{R}^{m \times m}$, and $V \in \mathbb{R}^{n \times m}$).

If the rank of $M \in \mathbb{R}^{m \times n}$ is r, there exist $Q \in \mathbb{R}^{m \times r}$ and $A \in \mathbb{R}^{r \times n}$ such that $M = QA$ and $Q^T Q = I$. In fact, since the rank is r, we can take as the basis each column Q_i ($i = 1, \ldots, r$) of Q, and the j-th column ($j = 1, \ldots, n$) of M can be written as $\sum_{i=1}^{r} Q_i a_{i,j}$.

Then, the optimum $A = (a_{i,j})$ can be written as $Q^T Z$ using Q. In fact, if we differentiate

$$\frac{1}{2}\|Z - QA\|_F^2 = \frac{1}{2} \sum_{i=1}^{m} \sum_{j=1}^{n} (z_{ij} - \sum_{k=1}^{r} q_{i,k} a_{k,j})^2$$

by the element $a_{p,q}$ of A, then it becomes $-\sum_{i=1}^{m}(z_{i,q} - \sum_{k=1}^{r} q_{i,k} a_{k,q}) q_{i,p}$, which is the (p, q)-th element of $-Q^T(Z - QA) = -Q^T Z + A$.

Moreover, for $\Sigma = ZZ^T$ and $M := QQ^T Z$, we have

$$\|Z - M\|_F^2 = \|Z^T(I - QQ^T)\|_F^2 = \text{trace } \Sigma - \text{trace}(Q^T \Sigma Q) \, ,$$

which means that the problem reduces to finding the value of Q that maximizes trace$(Q^T \Sigma Q)$. In fact, we derive from (6.1)

$$\|Z^T(I - QQ^T)\|_F^2 = \text{trace}((I - QQ^T)\Sigma(I - QQ^T)),$$

and the right-hand side can be transformed into

$$\text{trace}((I - QQ^T)^2\Sigma) = \text{trace}((I - QQ^T)\Sigma) = \text{trace} \, \Sigma - \text{trace}(Q^T\Sigma Q),$$

where (6.2) has been used for the first and last variables. Thus, the problem further reduces to maximizing trace$(Q^T \Sigma Q)$. Then, using $U^T U = UU^T = I$, $Q^T Q = I$ and $R := U^T Q \in \mathbb{R}^{m \times r}$, from

$$Q^T \Sigma Q = Q^T ZZ^T Q = Q^T UD^2 U^T Q = R^T D^2 R,$$

the problem reduces to maximizing trace$(R^T D^2 R)$ under $R^T R = Q^T UU^T Q = Q^T Q = I$.

Furthermore, if we let $H := RR^T \in \mathbb{R}^{m \times m}$ and write the (i, j)-th element as $h_{i,j}$, then, as we will see, $h_{1,1} = \cdots = h_{r,r} = 1$ and $h_{r+1,r+1} = \cdots = h_{m,m} = 0$ is the optimum solution.

If we apply (6.2), then we have

$$\text{trace} \, H = \text{trace}(RR^T) = \text{trace}(R^T R) = r.$$

From $h_{i,j} = \sum_{k=1}^r r_{i,k} r_{j,k}$, we have $h_{i,i} = \sum_{k=1}^r r_{i,k}^2 \geq 0$. Moreover, if we compare it with the $S \in \mathbb{R}^{m \times m}$ $S^T S = SS^T = I_m$, which is obtained by adding the $m - r$ columns to $R \in \mathbb{R}^{m \times r}$, then $h_{i,i}$ is the squared sum over the columns of R along the i-th row and does not exceed 1. Because the (i, j)-th element of $R^T D^2 R$ is $\sum_{k=1}^m r_{k,i} r_{k,j} d_k^2$, from (6.2), the trace is

$$\text{trace}(R^T D^2 R) = \text{trace}(RR^T D^2) = \sum_{i=1}^m h_{i,i} d_i^2$$

From $R^T R = I_r$, we have $\sum_{i=1}^m h_{i,i} = \sum_{i=1}^m \sum_{j=1}^r r_{i,j}^2 = r$. Thus, $h_{1,1} = \cdots = h_{r,r} = 1$ and $h_{r+1,r+1} = \cdots = h_{m,m} = 0$ is the solution.

Then, for $i = r + 1, \ldots, m$, $h_{i,i} = 0$ means $\sum_{j=1}^r r_{i,j}^2 = 0$, i.e., $r_{i,j} = 0$ ($j = 1, \ldots, r$). Therefore, this R can be written as $R = \begin{bmatrix} R_r \\ O \end{bmatrix} \in \mathbb{R}^{m \times r}$ using an arbitrary orthogonal matrix $R_r \in \mathbb{R}^{r \times r}$, where $O \in \mathbb{R}^{r \times (m-r)}$ is the zero matrix.

Note that the matrix U_r obtained by replacing U with zeros except the first r columns is the Q that corresponds to the solution. In fact, for $U = [U_r \; U_{m-r}]$,

$U_r \in \mathbb{R}^{m \times r}$, $U_{m-r} \in \mathbb{R}^{m \times (m-r)}$, and $Q \in \mathbb{R}^{m \times r}$, we have $\begin{bmatrix} R_r \\ O \end{bmatrix} = \begin{bmatrix} U_r^T \\ U_{m-r}^T \end{bmatrix} Q$ and

$Q = \begin{bmatrix} U_r & U_{m-r} \end{bmatrix} \begin{bmatrix} R_r \\ O \end{bmatrix} = U_r R_r$.

Finally, we have

$$M = QQ^T Z = U_r U_r^T Z = U_r U_r^T [U_r \; U_{m-r}] DV^T$$
$$= U_r [I \; O] DV^T = \begin{bmatrix} U_r & U_{m-r} \end{bmatrix} \begin{bmatrix} D_r & O \\ O & O \end{bmatrix} V^T = UD_r V^T ,$$

which completes the proof. $\square$

Proposition 20 *For the matrices* $Q, M \in \mathbb{R}^{m \times n}$, *if* $d_1(M), \ldots, d_n(M)$ *are the singular values of* M, *then we have*

$$\langle Q, M \rangle \leq \|Q\|^* \sum_{i=1}^n d_i(M) ,$$

where the equality holds if and only if

$$d_1(Q) = \cdots = d_r(Q) \geq d_{r+1}(Q) = \cdots = d_n(Q) = 0$$

and either $u_i(Q) = u_i(M)$, $v_i(Q) = v_i(M)$ *or* $u_i(Q) = -u_i(M)$, $v_i(Q) = -v_i(M)$ *for each* $i = 1, \ldots, r$, *where we singular-decomposed* M, Q *as* $M = \sum_{i=1}^n d_i(M)u_i(M)v_i(M)$ *and* $Q = \sum_{i=1}^n d_i(Q)u_i(Q)v_i(Q)$.

Proof Using (6.2), we have

$$\langle Q, M \rangle = \text{trace}(Q^T UDV^T) = \text{trace}(V^T Q^T UD) = \langle U^T QV, D \rangle$$
$$= \sum_{i=1}^n d_i(M) \cdot (U^T QV)_{i,i} = \sum_{i=1}^n d_i(M)u_i(M)^T Qv_i(M)$$
$$= \sum_{i=1}^n d_i(M)u_i(M)^T u_i(Q)d_i(Q)v_i^T(Q)v_i(M)$$
$$\leq \sum_i d_i(M)d_1(Q) = \|M\|_* \|Q\|^*$$

with the equality if and only if for i such that $d_i \neq 0$,

$$u_i(M)^T u_i(Q)d_i(Q)v_i^T(Q)v_i(M) = d_1(Q) ,$$

which is $d_i(Q) = d_1(q)$ and

$$u_i(M)^T u_i(Q) \cdot v_i(Q)^T v_i(M) = 1 \qquad (6.14)$$

for $i = 1, \ldots, r$, because the lengths of $u_i(M), u_i(Q), v_i(M), v_i(Q)$ are ones. Note that (6.14) means either $u_i(M) = u_i(Q), \ v_i(Q) = v_i(M)$ or $u_i(M) = -u_i(Q), v_i(Q) = -v_i(M)$, which completes the proof.   □

Exercises 76–87

76. For the singular decomposition, answer the following inquiries.

   (a) For the symmetric matrices, what relations exist between singular and eigen-value decompositions? How about the nonnegative definite matrices?
   (b) Suppose that $m \leq n$. We define the singular decomposition of $Z \in \mathbb{R}^{m \times n}$ as the transpose of the singular decomposition $Z^T = \bar{U} D \bar{V}^T$. What are the sizes of U, D, V in $Z = UDV^T$?

77. Show the following:

   (a) $\|Z\|_F^2 = \mathrm{trace}(Z^T Z)$, where $\|Z\|_F$ is the square root of the squared sum of the mn elements in $Z \in \mathbb{R}^{m \times n}$ の個 (Z の Frobenius norm).
      Hint For $Z = (z_{i,j})$, the i-th diagonal element of $Z^T Z$ is $\sum_{j=1}^{n} z_{i,j}^2$.
   (b) Let $l, m, n \geq 1$. For matrices $A \in \mathbb{R}^{l \times m}, \ B \in \mathbb{R}^{m \times n}, \ C \in \mathbb{R}^{n \times l}$, we have $\mathrm{trace}(ABC) = \mathrm{trace}(BCA)$.

78. Raise a counterexample against the statement "The matrix rank is a convex function":

   For arbitrary $A, B \in \mathbb{R}^{m \times n}$ and $0 < \alpha < 1$, we have

   $$\mathrm{rank}(\alpha A + (1 - \alpha)B) \leq \alpha \cdot \mathrm{rank}\, A + (1 - \alpha)\, \mathrm{rank}\, B .$$

   Hint Find A, B such that $m = n = 2$ rank $A = $ rank $B = 1$.

79. When the singular decomposition of $Z \in \mathbb{R}^{m \times n}$ can be written as $Z = UDV^T$,[1] we show that given $Z = UDV^T$, the matrix M that minimizes $\|Z - M\|_F$ and has a rank of at most r is $UD_r V^T$ (Eckart-Young's theorem), where D_r is the diagonal matrix D such that $d_{r+1} = \cdots = d_n = 0$. We assume that the rank of Z is m.

   (a) If the rank of M is r, using a matrix $Q \in \mathbb{R}^{m \times r}$ ($Q^T Q = I_r, m \geq r$), we can write it as $M = QA$. In fact, if the rank is r, we can take as basis the columns Q_j ($j = 1, \ldots, r$) of Q, which means that the j-th column of M can be written as $\sum_{i=1}^{r} Q_i a_{i,j}$. Show that the optimum $A = (a_{i,j})$ can be written as $Q^T Z$ using the matrix Q.
      Hint If we differentiate $\frac{1}{2}\|Z - QA\|_F^2 = \frac{1}{2} \sum_i \sum_j (z_{ij} - \sum_k q_{ik} a_{kj})^2$ by

---

[1] If $m \geq n$, then $U \in \mathbb{R}^{m \times n}, D \in \mathbb{R}^{n \times n}$, and $V \in \mathbb{R}^{n \times n}$, and if $m \leq n$, then $U \in \mathbb{R}^{m \times m}, D \in \mathbb{R}^{m \times m}$, and $V \in \mathbb{R}^{n \times m}$.

the element $a_{p,q}$ of A, we obtain $-\sum_{i=1}^{m}(z_{i,q} - \sum_{k=1}^{r} q_{i,k}a_{k,q})q_{i,p}$, which is the
(p, q)-th element of $-Q^T(Z - QA)$.

(b) Let $\Sigma := ZZ^T$. Show that

$$\|Z - M\|_F^2 = \|Z^T(I - QQ^T)\|_F^2 = \text{trace } \Sigma - \text{trace}(Q^T \Sigma Q) ,$$

which means that minimizing $\|Z - M\|_F^2$ reduces to finding the Q that max-
imizes $\text{trace}(Q^T \Sigma Q)$.
Hint $\|Z^T(I - QQ^T)\|_F^2 = \text{trace}((I - QQ^T)\Sigma(I - QQ^T))$ can be
derived from Exercise 77 (a) and becomes $\text{trace}((I - QQ^T)^2\Sigma)$ from
Exercise 77 (b).

(c) Show that (b) further reduces to finding the orthogonal matrix $R \in \mathbb{R}^{m \times r}$
that maximizes $\text{trace}(R^T D^2 R)$.
Hint Show that $Q^T \Sigma Q = Q^T U D^2 U^T Q$ and

$$Q \text{orthogonal} \iff R := U^T Q \text{orthogonal}.$$

(d) Let $H = RR^T$ and $h_{i,i}$ $(i = 1, \ldots, m)$ be the diagonal elements. By proving
the following statements, show that $h_{1,1} = \cdots = h_{r,r} = 1, h_{r+1,r+1} = \cdots =
h_{m,m} = 0$ is the optimum solution.
  i. $\sum_{i=1}^{m} h_{i,i} = r$
  ii. $\text{trace } H = r$
  iii. $0 \le h_{i,i} \le 1$
  iv. $\text{trace}(R^T D^2 R) = \sum_{i=1}^{m} h_{i,i} d_i^2$
    Hint Because R becomes a square orthogonal matrix by adding
    columns, for the square matrix, the squared row and column sums are
    both one.

(e) Show that the solution of (d) can be written as $R = \begin{bmatrix} R_r \\ O \end{bmatrix} \in \mathbb{R}^{m \times r}$ using
an arbitrary orthogonal matrix $R_r \in \mathbb{R}^{r \times r}$, where $O \in \mathbb{R}^{r \times (m-r)}$ is the zero
matrix.

(f) Let U_+ be the matrix U with zeros in all columns except the first r columns.
Show that U_+ is the Q that corresponds to (e) and that $M = QQ^T Z =
U_+ U_+^T Z = U D_r V^T$.
Hint For $U = [U_+ \ U_-], U_+ \in \mathbb{R}^{m \times r}$, and $U_- \in \mathbb{R}^{m \times (m-r)}, Q \in \mathbb{R}^{m \times r}$, we
have $\begin{bmatrix} R_r \\ O \end{bmatrix} = \begin{bmatrix} U_+^T \\ U_-^T \end{bmatrix} Q$, from which we derive $Q = U_+ R_r$.

80. Based on Exercise 79, we construct a function svd.r to approximate the matrix
z by the rank r. Fill in the blanks.
Hint Transpose ss$v before multiplication.

```
1  svd.r = function(z, r) {
2    n = min(nrow(z), ncol(z)); ss = svd(z)
```

```
3    return(ss$u %*% diag(c(ss$d[1:r], rep(0, n - r))) %*% ## Blank
         (1) ##)
4  }
5  ## if r is assume to be at least two, the following is ok.
6  svd.r = function(z, r) {
7    ss = svd(z); return(ss$u[, 1:r] %*% diag(ss$d[1:r]) %*% ## Blank
         (2) ##)
8  }
```

Moreover, execute the following to examine the behavior.

```
1  m = 100; n = 80; z = matrix(rnorm(m * n), nrow = m)
2  F.norm = NULL
3  for (r in 1:n) {m = svd.r(z, r); F.norm = c(F.norm, norm(z - m, "F"
       ))}
4  plot(1:n, F.norm, type = "l", xlab = "Rank", ylab = "Squared
       Frobenius norm")
```

81. The following code is the process of obtaining another image file with the same lower rank as that of lion.jpg. Generally, in the JPG format, each pixel has a 256 grayscale of information. It has information about the three colors blue, red, and green. For example, if we have 480×360 pixels vertically and horizontally, each pixel of $480 \times 360 \times 3$ will hold one of 256 values. In the code below, the rank is $r = 2, 5, 10, 20, 50, 100$. This time, we approximate the lower ranks for red, green, and blue (Fig. 6.2). It is necessary to create a folder called /compressed under the current folder in advance.

```
1  library(jpeg)
2  image = readJPEG('lion.jpg')
3  rank.seq = c(2, 5, 10, 20, 50, 100)
4  mat = array(0, dim = c(nrow(image), ncol(image), 3))
5  for (j in rank.seq) {
6    for (i in 1:3) mat[, , i] = svd.r(image[, , i], j)
7    writeJPEG(mat,
8             paste("compressed/lion_compressed", "_svd_rank_", j, ".
         jpg", sep = ""))
9  }
```

82. Suppose we observe only the position of $\Omega \subseteq \{1, \ldots, m\} \times \{1, \ldots, n\}$ in the matrix $Z \in \mathbb{R}^{m \times n}$. From $z_{i,j}$ $((i, j) \in \Omega)$, we find the matrix M of rank r that minimizes the Frobenius norm of the difference between the matrices M and Z. In the following process, mask is a matrix with 1 and 0 at the observed and unobserved positions, respectively (in $1 - $ mask, the values $1, 0$ are reversed). In the inner loop, the rank is approximated to r, and the unobserved value is predicted ten times. In the outer loop, to prevent falling into a local solution, the initial value is changed and executed five times to find the solution that minimizes the Frobenius norm of the difference. Fill in the blanks, and execute the process.

```
1  mat.r = function(z, mask, r) {
2    z = as.matrix(z)
3    min = Inf
```

```
4    m = nrow(z); n = ncol(z)
5    for (j in 1:5) {
6       guess = matrix(rnorm(m * n), nrow = m)
7       for (i in 1:10) guess = svd.r(mask * z + (1 - mask) * guess, r)
8       value = norm(mask * (z - guess), "F")
9       if (value < min) {min.mat = ## Blank (1) ##; min = ## Blank (2) ##}
10   }
11   return(min.mat)
12 }
13
14 library(jpeg)
15 image = readJPEG('lion.jpg')
16 m = nrow(image); n = ncol(image)
17 mask = matrix(rbinom(m * n, 1, 0.5), nrow = m)
18 rank.seq = c(2, 5, 10, 20, 50, 100)
19 mat = array(0, dim = c(nrow(image), ncol(image), 3))
20 for (j in rank.seq) {
21   for (i in 1:3) mat[, , i] = mat.r(image[, , i], mask, j)
22   writeJPEG(mat,
23             paste("compressed/lion_compressed", "_mat_rank_", j, ".jpg"
              , sep = ""))
24 }
```

In the following, when the singular values of Z are $d_1, \ldots, d_n$, we consider the sum $\sum_{i=1}^{n} d_i(Z)$ to be the nuclear norm and write it as $\|Z\|_*$. Moreover, we consider the maximum $\max\{d_1, \ldots, d_n\}$ to be the spectral norm and write it as $\|Z\|^*$.

83. Prove the following on norms.

   (a) The matrix norms are convex.
   (b) The dual norms satisfy the definition of norms.
   (c) The nuclear norm is the dual of the spectral norm.

84. For matrices A, B with the same size, show $\|A\|_*\|B\|^* \geq \langle A, B \rangle$, and derive the condition under which the equality holds.

85. Derive the subderivative of the nuclear norm $\|\cdot\|_*$ at matrix M.

86. For $Z \in \mathbb{R}^{m \times n}$, $\Omega \subseteq \{(i, j) \mid i = 1, \ldots, m, \ j = 1, \ldots, n\}$, $\lambda > 0$, we wish to find the $M \in \mathbb{R}^{m \times n}$ that minimizes

$$L := \frac{1}{2} \sum_{(i,j) \in \Omega} (z_{i,j} - m_{i,j})^2 + \lambda \|M\|_* \quad \text{(cf. (6.11))}. \qquad (6.15)$$

Suppose that we have observations $Z \in \mathbb{R}^{m \times n}$ only for the positions $\Omega \subseteq \{1, \ldots, m\} \times \{1, \ldots, n\}$ and that we take the subderivative of (6.11) w.r.t. $M = (m_{i,j})$ to obtain the following equation:

$$[\Omega, M - Z, 0] + \lambda \left(\sum_{i=1}^{r} u_i v_i^T + \sum_{k=r+1}^{n} d_k \tilde{u}_k \tilde{v}_k^T \right) = 0. \qquad (6.16)$$

Mazumder et al. (2010) [19] showed that starting with an arbitrary initial value M_0 and updating M_t via

$$M_{t+1} \leftarrow S_\lambda([\Omega, Z, M_t]) \qquad (6.17)$$

until convergence, both sides of M in (6.17) coincide. Show that this matrix M satisfies (6.16), where $S_\lambda(W)$ denotes the matrix whose singular values d_i, $i = 1, \ldots, n$, are replaced with $d_i - \lambda$ if $d_i \geq \lambda$ and zeros otherwise.

Hint Singular-decompose $[\Omega, Z, M]$, and subtract $\min\{\lambda, d_k\}$ from the singular values d_k such that the obtained singular values are nonnegative. If we set $M_{t+1} = M_t = M$, then we obtain (6.16).

87. We construct the function `soft.svd` that returns $Z' = UD'V^T$ given $Z = UDV^T$, where D and D' are diagonal matrices with the diagonal elements d_k and $d'_k := \max\{d_k - \lambda, 0\}, k = 1, \ldots .n$. Fill in the blank.

```
1  soft.svd = function(lambda, z) {
2    n = ncol(z); ss = svd(z); dd = pmax(ss$d - lambda, 0)
3    return(## Blank ##)
4  }
```

For the program below, obtain three images, changing the values of p and λ.

```
1  mat.lasso = function(lambda, z, mask) {
2    z = as.matrix(z); m = nrow(z); n = ncol(z)
3    guess = matrix(rnorm(m * n), nrow = m)
4    for (i in 1:20) guess = soft.svd(lambda, mask * z + (1 - mask) *
         guess)
5    return(guess)
6  }
7
8  library(jpeg)
9  image = readJPEG('lion.jpg')
10 m = nrow(image[, , 1]); n = ncol(image[, , 1])
11 p = 0.5
12 lambda = 0.5
13 mat = array(0, dim = c(m, n, 3))
14 mask = matrix(rbinom(m * n, 1, p), ncol = n)
15 for (i in 1:3) mat[, , i] = mat.lasso(lambda, image[, , i], mask)
16 writeJPEG(mat, paste("compressed/lion_compressed", "_mat_soft.jpg",
         sep = ""))
```

Chapter 7
Multivariate Analysis

In this chapter, we consider sparse estimation for the problems of multivariate analysis, such as principal component analysis and clustering. There are two equivalence definitions for principal component analysis: finding orthogonal vectors that maximize the variance and finding a vector that minimizes the reconstruction error when the dimension is reduced. This chapter first introduces the sparse estimation methods for principal component analysis, of which SCoTLASS and SPCA are popular. In each case, the purpose is to find a principal component vector with few nonzero components. On the other hand, introducing sparse estimation is crucial for clustering to select variables. In particular, we consider the K-means and convex clustering problems. The latter has the advantage of not falling into a locally optimal solution because it becomes a problem of convex optimization.

7.1 Principal Component Analysis (1): SCoTLASS

In the following, we assume that the matrix $X = (x_{i,j}) \in \mathbb{R}^{N \times p}$ is centered: we subtract the arithmetic mean from each column, and the average in each column is zero $\sum_{i=1}^{N} x_{i,j} = 0$ $(j = 1, \ldots, p)$.

Let v_1 be the $v \in \mathbb{R}^p$ for $\|v\| = 1$ that maximizes

$$\|Xv\|^2 . \tag{7.1}$$

Let v_2 be the $v \in \mathbb{R}^p$ for $\|v\| = 1$ that is orthogonal to v_1 and maximizes (7.1), and so on. In this way, finding the orthonormal system $V = [v_1, \ldots, v_p]$ is called principal component analysis (PCA) . We consider the constraint where $v_1, \ldots, v_p$ are orthogonal later and first find the v that maximizes $\|Xv\|^2$ under $\|v\| = 1$. For each $j = 1, \ldots, p$, because this v_j maximizes

© The Author(s), under exclusive license to Springer Nature Singapore Pte Ltd. 2021
J. Suzuki, *Sparse Estimation with Math and R*,
https://doi.org/10.1007/978-981-16-1446-0_7

$$\|Xv_j\|^2 - \mu(\|v_j\|^2 - 1) ,$$

if we differentiate by v_j, we have

$$X^T X v_j - \mu_j v_j = 0$$

More precisely, using the sample covariance $\Sigma := \dfrac{1}{N} X^T X$ based on X and $\lambda_j := \dfrac{\mu_j}{N}$, we can express

$$\Sigma v_j = \lambda_j v_j . \tag{7.2}$$

We note that each column V is a basis of the eigenspace Σ. In case some $\lambda_1 \geq \cdots \geq \lambda_p$ values conflict, we choose the eigenvectors such that they are orthogonal. Because Σ is nonnegative definite, the eigenvectors with different eigenvalues are orthogonal. If all of the eigenvalues are different, i.e., $\lambda_1 > \cdots > \lambda_p$, then $v_1, \ldots, v_p$ are orthogonal.

In reality, we use only the first m ($1 \leq m \leq p$) rather than all $v_1, \ldots, v_p$. If we project each row of X to $V_m := [v_1, \ldots, v_m] \in \mathbb{R}^{p \times m}$, we obtain $Z := X V_m \in \mathbb{R}^{N \times m}$. We project the p-dimensional information to the space of the m principle components $v_1, \ldots, v_m$ to approximate the p-dimensional space X by the m-dimensional Z. The PCA is the linear map for this compression.

Although there are several approaches to sparse estimation for PCA, we first consider restricting the number of zero elements in PCA.

When we find the $v \in \mathbb{R}^p$ for $\|v\|_2 = 1$ that maximizes $\|Xv\|_2$, if we restrict the number $\|v\|_0$ of nonzero elements in v, then the formulation maximizing

$$v^T X^T X v - \lambda \|v\|_0$$

under $\|v\|_0 \leq t, \|v\|_2 = 1$, for some integer t is not convex. On the other hand, when we wish to add the constraint $\|v\|_1 \leq t$ ($t > 0$), the formulation maximizing

$$v^T X^T X v - \lambda \|v\|_1 \tag{7.3}$$

under $\|v\|_2 = 1$ is not convex.

The formulation maximizing

$$u^T X v - \lambda \|v\|_1 \tag{7.4}$$

under $\|u\|_2 = \|v\|_2 = 1$ for $u \in \mathbb{R}^N$ is proposed (SCoTLASS) [15] (simplified component technique-Lasso). The optimum v obtained in (7.4) is optimized in (7.3) as well. In fact, differentiating

$$L := -u^T X v + \lambda \|v\|_1 + \frac{\mu}{2}(u^T u - 1) + \frac{\delta}{2}(v^T v - 1) \tag{7.5}$$

by u, from $Xv - \mu u = 0$, $\|u\| = 1$, we have $u = \dfrac{Xv}{\|Xv\|_2}$. If we substitute this into (7.5), we obtain

$$-\|Xv\|_2 + \lambda\|v\|_1 + \frac{\delta}{2}(v^T v - 1).$$

Although (7.5) takes positive values when we differentiate it by each of u, v twice, it is not convex w.r.t. (u, v).

Example 60 When $N = p = 1$, $X > \sqrt{\mu\delta}$, it is not convex. In fact, because L is a bivariate function of u, v, we have

$$\nabla^2 L = \begin{bmatrix} \dfrac{\partial^2 L}{\partial u^2} & \dfrac{\partial^2 L}{\partial u \partial v} \\ \dfrac{\partial^2 L}{\partial u \partial v} & \dfrac{\partial^2 L}{\partial v^2} \end{bmatrix} = \begin{bmatrix} \mu & -X \\ -X & \delta \end{bmatrix}.$$

If the determinant $\mu\delta - X^2$ is negative, $\nabla^2 L$ contains a negative eigenvalue.

Thus, if $\alpha \mapsto f(\alpha, \beta)$ and $\beta \mapsto f(\alpha, \beta)$ are convex w.r.t. $\alpha \in \mathbb{R}^m$ and $\beta \in \mathbb{R}^n$, respectively, then we say that $f : \mathbb{R}^m \times \mathbb{R}^n \to \mathbb{R}$ is biconvex. In general, convexity does not mean convexity.

The reason why we transform (7.4) is that we can efficiently obtain the solution of (7.3): Choose an arbitrary $v \in \mathbb{R}^p$ such that $\|v\|_2 = 1$. Repeat to update $u \in \mathbb{R}^N$, $v \in \mathbb{R}^p$:

1. $u \leftarrow \dfrac{Xv}{\|Xv\|_2}$

2. $v \leftarrow \dfrac{S_\lambda(X^T u)}{\|S_\lambda(X^T u)\|_2}$

until the u, v values converge, where $S_\lambda(z)$ is the function $S_\lambda(\cdot)$ introduced in Chap. 1 that applies each element of $z = [z_1, \cdots, z_p] \in \mathbb{R}^p$, in which $v = \dfrac{S_\lambda(X^T u)}{\|S_\lambda(X^T u)\|_2}$ and $u = \dfrac{Xv}{\|Xv\|_2}$ are obtained from $\dfrac{\partial L}{\partial u} = \dfrac{\partial L}{\partial v} = 0$.

In fact, if we take the subderivative of $\|v\|_1$, we have $1, -1, [-1, 1]$ for $v_j > 0$, $v_j < 0$, $v_j = 0$, respectively, for $j = 1, \ldots, p$, which means that

$$\begin{cases} \text{(the } j - \text{th column of } X^T u) - \lambda + \delta v_j = 0, & \text{(the } j - \text{th column of } X^T u) > \lambda \\ \text{(the } j - \text{th column of } X^T u) + \lambda + \delta v_j = 0, & \text{(the } j - \text{th column of } X^T u) < -\lambda \\ \text{(the } j - \text{th column of } X^T u) + \lambda[-1, 1] \ni 0, & -\lambda \leq \text{(the } j - \text{th column of } X^T u) \leq \lambda \end{cases},$$

i.e., $\delta v = -S_\lambda(X^T u)$.

In general, when $\Psi : \mathbb{R}^p \times \mathbb{R}^p \to \mathbb{R}$ satisfies

$$\begin{cases} f(\beta) \leq \Psi(\beta, \theta), & \theta \in \mathbb{R}^p \\ f(\beta) = \Psi(\beta, \beta) \end{cases} \tag{7.6}$$

for $f : \mathbb{R}^p \to \mathbb{R}$, we say that Ψ majorizes f at a point $\beta \in \mathbb{R}^p$.

When Ψ majorizes f, if we arbitrarily choose $\beta^0 \in \mathbb{R}^p$ and generate $\beta^1, \beta^2, \dots$ via the recurrence formula

$$\beta^{t+1} = \arg\min_{\beta \in \mathbb{R}^p} \Psi(\beta, \beta^t) ,$$

we have

$$f(\beta^t) = \Psi(\beta^t, \beta^t) \geq \Psi(\beta^{t+1}, \beta^t) \geq f(\beta^{t+1}) .$$

In SCoTLASS, for

$$f(v) := -\|Xv\|_2 + \lambda\|v\|_1$$

$$\Psi(v, v') := -\frac{(Xv)^T(Xv')}{\|Xv'\|_2} + \lambda\|v\|_1 ,$$

Ψ majorizes f at v. In fact, we check (7.6) via the Schwartz inequality.

After arbitrarily giving v^0, if we generate $v^0, v^1, \dots$ via

$$v^{t+1} := \arg\max_{\|v\|_2=1} \Psi(v, v') ,$$

then each value expresses $\dfrac{\mathcal{S}_\lambda(X^T u)}{\|\mathcal{S}_\lambda(X^T u)\|_2}$ at the time, and it is monotonically decreasing.

For example, we can construct the function SCoTLASS as follows.

```
1  soft.th = function(lambda, z) return(sign(z) * pmax(abs(z) - lambda, 0))
2  ## even if z is a vector, soft.th works
3  SCoTLASS = function(lambda, X) {
4    n = nrow(X); p = ncol(X); v = rnorm(p); v = v / norm(v, "2")
5    for (k in 1:200) {
6      u = X %*% v; u = u / norm(u, "2"); v = t(X) %*% u
7      v = soft.th(lambda, v); size = norm(v, "2")
8      if (size > 0) v = v / size else break
9    }
10   if (norm(v, "2") == 0) print("all the elements of v are zero"); return(
       v)
11 }
```

Example 61 Due to nonconvexity, even when λ increases, $\|Xv\|_2$ may not be monotonically decreasing even though the number of zeros in v decreases with λ. We show the execution of the code below in Fig. 7.1.

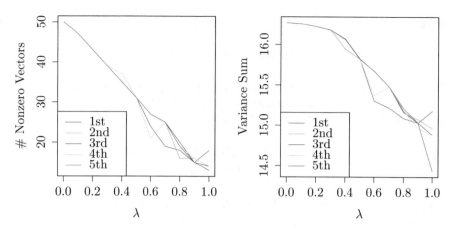

Fig. 7.1 Execution of Example 61. With λ, the number of nonzero vectors and the sum of the variances decrease monotonically, If we change the initial value and execute it, these values will be different each time. This tendency becomes more remarkable as the value of λ increases

```
1   ## Data Generation
2   n = 100; p = 50; X = matrix(rnorm(n * p), nrow = n); lambda.seq = 0:10
        / 10
3
4   m = 5; SS = array(dim = c(m, 11)); TT = array(dim = c(m, 11))
5   for (j in 1:m) {
6     S = NULL; T = NULL
7     for (lambda in lambda.seq) {
8       v = SCoTLASS(lambda, X); S = c(S, sum(sign(v ^ 2))); T = c(T, norm
          (X %*% v, "2"))
9     }
10    SS[j, ] = S; TT[j, ] = T
11  }
12  ## Display
13  par(mfrow = c(1, 2))
14  SS.min = min(SS); SS.max = max(SS)
15  plot(lambda.seq, xlim = c(0, 1), ylim = c(SS.min, SS.max),
16       xlab = "lambda", ylab = "# of nonzero vectors")
17  for (j in 1:m) lines(lambda.seq, SS[j, ], col = j + 1)
18  legend("bottomleft", paste0(1:5, "-th"), lwd = 1, col = 2:(m + 1))
19  TT.min = min(TT); TT.max = max(TT)
20  plot(lambda.seq, xlim = c(0, 1), ylim = c(TT.min, TT.max), xlab = "
        lambda", ylab = "Variance Sum")
21  for (j in 1:m) lines(lambda.seq, TT[j, ], col = j + 1)
22  legend("bottomleft", paste0(1:5, "-th"), lwd = 1, col = 2:(m + 1))
23  par(mfrow = c(1, 1))
```

To handle not only the first principal component but also multiple components, the following formulation is made in SCoTLASS. We formulate maximizing

$$u_k^T X v_k$$

under

$$\|v_k\|_2 \le 1 \,, \ \|v_k\|_1 \le c \,, \ \|u_k\|_2 \le 1 \,, \ u_k^T u_j = 0 \,, \ (j = 1, \ldots, k-1)$$

for $k = 1, \ldots, m$, given $c > 0$. In fact, if we include sparsity in the upstream eigen-vector settings, there is no guarantee that it will be orthogonal to the eigenvectors downstream. The u_k when v_k is fixed is given by

$$u_k := \frac{P_{k-1}^{\perp} X v_k}{\|P_{k-1}^{\perp} X v_k\|_2} \tag{7.7}$$

for $P_{k-1}^{\perp} := I - \sum_{i=1}^{k-1} u_i u_i^T$. In fact, given $u_1, \ldots, u_{k-1}$, we have

$$u_j^T P_{k-1}^{\perp} X v_k = u_j^T (I - \sum_{i=1}^{k-1} u_i u_i^T) X v_k = 0 \quad (j = 1, \ldots, k-1) \,.$$

The u_k that maximizes $L = u_k^T X v_k - \mu(u_k^T u_k - 1)$ is (Eq. 7-6). Compared with the $m = 1$ case, from (7.7), the difference is only in multiplying $P_{k-1}^{\perp}$, where $P_{k-1}^{\perp}$ is the unit matrix for $k = 1$.

7.2   Principle Component Analysis (2): SPCA

If $m = p$, then V is nonsingular, which means that X can be recovered from $Z = XV$ via $ZV^{-1} = ZV^T = X$. In general, however, because we have $XV_m V_m^T \ne X$, $XV_m V_m^T$ and X do not coincide. Because each sample $x_i \in \mathbb{R}^p$ (the i-th row vector in X) is transformed into $x_i V_m$, we obtain $x_i V_m V_m^T$ when we recover, and the difference is $x_i(I - V_m V_m^T) \in \mathbb{R}^p$. We define the reconstruction error by

$$\sum_{i=1}^{N} \|x_i(I - V_m V_m^T)\|^2 = \sum_{i=1}^{N} x_i(I - V_m V_m^T)^2 x_i^T = \sum_{i=1}^{N} x_i(I - V_m V_m^T) x_i^T \,. \tag{7.8}$$

Then, given X, (7.8) is minimized when

$$\sum_{i=1}^{N} x_i V_m V_m^T x_i^T = \text{trace}(X V_m V_m^T X^T) = \text{trace}(V_m^T X^T X V_m)$$

$$= \sum_{j=1}^{m} v_j^T X^T X v_j = \sum_{j=1}^{m} \|X v_j\|^2 \tag{7.9}$$

is maximized. Thus, PCA can be regarded as finding the $v_1, \ldots, v_m$ that minimize the reconstruction error. In fact, maximizing (7.9) under $\|v_1\|^2 = \cdots = \|v_m\|^2 = 1$ is maximizing

$$\sum_{j=1}^{m} \|Xv_j\|^2 - \sum_{j=1}^{m} \lambda_j(\|v_j\|^2 - 1) .$$

If we differentiate by v_j, we obtain (7.2).

SPCA (sparse principal component analysis) [32] is another formulation of PCA different from SCoTLASS introduced in the previous section. Note that the problem of minimizing

$$\min_{u,v\in\mathbb{R}^p, \|u\|_2=1} \left\{ \frac{1}{N} \sum_{i=1}^{N} \|x_i - x_i vu^T\|_2^2 + \lambda_1\|v\|_1 + \lambda_2\|v\|_2^2 \right\} \qquad (7.10)$$

is biconvex for u, v, where x_i is the i-th row vector of X. In fact, if we consider the constraint $\|u\|_2 = 1$, we can formulate (7.10) as

$$L := \frac{1}{N} \sum_{i=1}^{N} x_i x_i^T - \frac{2}{N} \sum_{i=1}^{N} x_i vu^T x_i^T + \frac{1}{N} \sum_{i=1}^{N} x_i vv^T x_i^T + \lambda_1\|v\|_1 + \lambda_2\|v\|_2^2 + \mu(u^T u - 1) .$$

We observe that if we differentiate it by u_j and v_k twice, then the results are nonnegative when excluding the term $\|v\|_1$.

When we optimize w.r.t. u when v is fixed, the solution is

$$u = \frac{X^T z}{\|X^T z\|_2}$$

with $z = (z_i)$ $(i = 1, \ldots, N)$, $z_1 = x_1 v, \ldots, z_N = x_N v$. In fact, if we differentiate L by u_j, the vector of the differentiated values for $j = 1, \ldots, p$ satisfies

$$-\frac{2}{N} \sum_{i=1}^{N} x_i vx_i^T + 2\mu u = 0.$$

When we optimize w.r.t. v when u is fixed, an extended algorithm of the elastic net can be applied. However, the problem is not convex (only biconvex), and we need to execute it several times, changing the initial value to obtain a solution close to the optimum.

If we take the subderivative of (7.10) w.r.t. v_k with u fixed, then when the constraint $\|v\|^2 = 1$ is missing, we have

$$\begin{cases} -\dfrac{1}{N}\displaystyle\sum_{i=1}^{N}\sum_{j=1}^{k-1} u_j x_{i,k}(r_{i,j,k} - u_j x_{i,k}v_k) + \lambda_1, & \dfrac{1}{N}\displaystyle\sum_{i=1}^{N}\sum_{j=1}^{k-1} r_{i,j}x_{i,k}u_j < -\lambda_1 \\[2em] -\dfrac{1}{N}\displaystyle\sum_{i=1}^{N}\sum_{j=1}^{k-1} u_j x_{i,k}(r_{i,j,k} - u_j x_{i,k}v_k) - \lambda_1, & \dfrac{1}{N}\displaystyle\sum_{i=1}^{N}\sum_{j=1}^{k-1} r_{i,j}x_{i,k}u_j > \lambda_1 \\[2em] 0, & \dfrac{1}{N}\displaystyle\sum_{i=1}^{N}\sum_{j=1}^{k-1} r_{i,j}x_{i,k}u_j < \lambda_1 \end{cases} ,$$

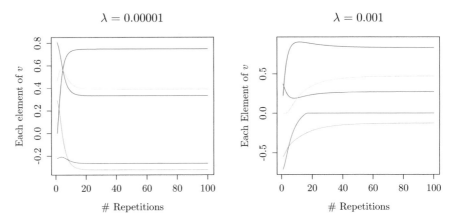

Fig. 7.2 Execution of Example 62. We observed the changes of v for $\lambda = 0.00001$(Left) and $\lambda = 0.001$(Right)

where $r_{i,j,k} = x_{i,j} - u_j \sum_{h \neq k} x_{i,h} v_h$. Thus, v_k is the normalized constraint such that $\|v\|_2 = 1$, i.e.,

$$
v_k = \frac{S_\lambda \left(\dfrac{1}{N} \sum_{i=1}^{N} x_{i,k} \sum_{j=1}^{k-1} r_{i,j,k} u_j \right)}{\sqrt{\sum_{h=1}^{p} S_\lambda \left(\dfrac{1}{N} \sum_{i=1}^{N} x_{i,h} \sum_{j=1}^{k-1} r_{i,j,h} u_j \right)^2}} \quad (k = 1, \ldots, p) \ .
$$

Example 62 We construct the SPCA process in the R language. We change the initial value several times and execute with each. Due to nonconvexity, we observe that when we increase the λ value, the set of nonzero variables does not monotonically grow. By repeating the process, we observe how each element of the vector v changes with the number of iterations (Fig. 7.2).

```
1   ## Data Generation
2   n = 100; p = 5; x = matrix(rnorm(n * p), ncol = p)
3   ## Computation of u,v
4   lambda = 0.001; m = 100
5   g = array(dim = c(m, p))
6   for (j in 1:p) x[, j] = x[, j] - mean(x[, j])
7   for (j in 1:p) x[, j] = x[, j] / sqrt(sum(x[, j] ^ 2))
8   r = rep(0, n)
9   v = rnorm(p)
10  for (h in 1:m) {
11      z = x %*% v
12      u = as.vector(t(x) %*% z)
13      if (sum(u ^ 2) > 0.00001) u = u / sqrt(sum(u ^ 2))
```

```
14   for (k in 1:p) {
15      for (i in 1:n) r[i] = sum(u * x[i, ]) - sum(u ^ 2) * sum(x[i, -k]
        * v[-k])
16      S = sum(x[, k] * r) / n
17      v[k] = soft.th(lambda, S)
18   }
19   if (sum(v ^ 2) > 0.00001) v = v / sqrt(sum(v ^ 2))
20   g[h, ] = v
21 }
22 ## Graph Display
23 g.max = max(g); g.min = min(g)
24 plot(1:m, ylim = c(g.min, g.max), type = "n",
25      xlab = "# Repetition", ylab = "Each element of v", main = "lambda
        = 0.001")
26 for (j in 1:p) lines(1:m, g[, j], col = j + 1)
```

When we consider sparse PCA not just for the first element but also for up to the m-th element, we minimize

$$L := \frac{1}{N} \sum_{i=1}^{N} \|x_i - x_i V_m U_m^T\|_2^2 + \lambda_1 \sum_{j=1}^{m} \|v_j\|_1 + \lambda_2 \sum_{j=1}^{m} \|v_j\|_2 + \mu \sum_{j=1}^{m} (u_j^T u_j - 1)$$

Compared with SCoTLASS, no orthogonality condition $u_j^T u_k = 0$ ($j = 1, \ldots, k - 1$) is included. Moreover, the formulation considers sparsity in minimizing the reconstruction error (7.8).

However, the formulations of SCoTLASS and SPCA are not convex but merely biconvex. SCoTLASS claims that the objective function decreases with the number of iterations, while SPCA claims the merit, and the elastic net algorithm is available for either of the two update formulas.

7.3 K-Means Clustering

We consider the problem of finding disjoint subsets $C_1, \ldots, C_K$ of $\{1, \ldots, N\}$ that minimize

$$\sum_{k=1}^{K} \sum_{i \in C_k} \|x_i - \bar{x}_k\|_2^2$$

from the data $x_1, \ldots, x_N \in \mathbb{R}^p$ and positive integer K to be K-means clustering, where $\bar{x}_k$ is the arithmetic mean of the data in C_k. Witten-Tibshirani (2010) formulated maximizing the weighted sum $\sum_{j=1}^{p} w_j a_j$ of p elements with

$$a_j(C_1, \ldots, C_K) := \frac{1}{N} \sum_{i=1}^{N} \sum_{i'=1}^{N} (x_{i,j} - x_{i',j})^2 - \sum_{k=1}^{K} \frac{1}{N_k} \sum_{i \in C_k} \sum_{i' \in C_k} (x_{i,j} - x_{i',j})^2$$

$$(7.11)$$

under the constraint $\|w\|_2 \le 1$, $\|w\|_1 \le s$, $w \ge 0$ (all the elements are nonnegative) for $s > 0$. The problem is to remove unnecessary variables by weighting for $j = 1, \ldots, p$ and to make the interpretation of the clustering easier. To solve the problem, they proposed repeating the following updates [31].

1. fixing $w_1, \ldots, w_p$, find the $C_1, \ldots, C_K$ that maximize $\sum_{j=1}^{p} w_j a_j(C_1, \ldots, C_K)$
2. fixing $C_1, \ldots, C_K$, find the $w_1, \ldots, w_p$ that maximize $\sum_{j=1}^{p} w_j a_j(C_1, \ldots, C_K)$

The first term in (7.11) is constant, and the second term can be written as

$$\frac{1}{N_k} \sum_{i \in C_k} \sum_{i' \in C_k} (x_{i,j} - x_{i',j})^2 = 2 \sum_{i \in C_k} (x_i - \bar{x}_k)^2$$

[1]; thus, to obtain the optimum $C_1, \ldots, C_N$ when the weights $w_1, \ldots, w_p$ are fixed, we may execute the following procedure.

```
1   k.means = function(X, K, weights = w) {
2     n = nrow(X); p = ncol(X)
3     y = sample(1:K, n, replace = TRUE); center = array(dim = c(K, p))
4     for (h in 1:10) {
5       for (k in 1:K) {
6         if (sum(y[] == k) == 0) center[k, ] = Inf else
7           for (j in 1:p) center[k, j] = mean(X[y[] == k, j])
8       }
9       for (i in 1:n) {
10        S.min = Inf
11        for (k in 1:K) {
12          if (center[k, 1] == Inf) break
13          S = sum((X[i, ] - center[k, ]) ^ 2 * w)
14          if (S < S.min) {S.min = S; y[i] = k}
15        }
16      }
17    }
18    return(y)
19  }
```

Example 63 Generate data artificially ($N = 1000$, $p = 2$), and execute the function k.means for the weights 1 : 1 and 1 : 100 (Fig. 7.3).

```
1   ## Data Generation
2   K = 10; p = 2; n = 1000; X = matrix(rnorm(p * n), nrow = n, ncol = p)
3   w = c(1, 1); y = k.means(X, K, w)
4   ## Display Output
5   plot(-3:3, -3:3, xlab = "x", ylab = "y", type = "n")
6   points(X[, 1], X[, 2], col = y + 1)
```

---

[1] See Chap. 10 of Statistical Learning with Math and R, Springer (2020).

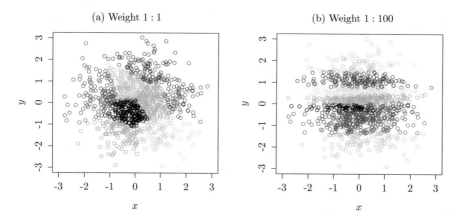

Fig. 7.3 We execute the function k.means for the weights 1 : 1 (left) and 1 : 100 (right). For case on the right whose weight of the second element is large, due to the larger penalty, the generated cluster is horizontally long

Comparing them, if we make the second element large, then we tend to obtain horizontal clusters.

Given $C_1, \ldots, C_K$, to obtain $w_1, \ldots, w_p$ that maximize (7.11) and $a \in \mathbb{R}^p$ with nonnegative elements, it is sufficient to find the value of w that maximizes the inner product $\sum_{h=1}^{p} w_h a_h$ under $\|w\|_2 = 1$, $\|w\|_1 \leq s$, $w \geq 0$.

Proposition 23 *Let $s > 0$ and $a \in \mathbb{R}^p$ with nonnegative elements. Then, the w with $\|w\|_2 = 1$, $\|w\|_1 \leq s$ that maximizes $w^T a$ can be written as*

$$w = \frac{\mathcal{S}_\lambda(a)}{\|\mathcal{S}_\lambda(a)\|_2} \tag{7.12}$$

for some $\lambda \geq 0$.

For the proof, see the Appendix.

More precisely, we can construct the following procedure [31]. In the function w.a, we obtain such a λ via binary search. At the beginning, we set $\lambda = \max_i a_i/2$ and $\delta = \lambda/2$. For each time, we divide δ by half, and if $\|w\|_1 > s$, add we δ to λ to make $\|w\|$ smaller, while if $\|w\|_1 < s$, we subtract δ from λ to make $\|w\|$ larger.

```
sparse.k.means = function(X, K, s) {
  p = ncol(X); w = rep(1, p)
  for (h in 1:10) {
    y = k.means(X, K, w)
    a = comp.a(X, y)
    w = w.a(a, s)
  }
  return(list(w = w, y = y))
}
```

```
10
11   w.a = function(a, s) {
12     w = rep(1, p)
13     a = a / sqrt(sum(a ^ 2))
14     if (sum(a) < s) return(a)
15     p = length(a)
16     lambda = max(a) / 2
17     delta = lambda / 2
18     for (h in 1:10) {
19       for (j in 1:p) w[j] = soft.th(lambda, a[j])
20       ww = sqrt(sum(w ^ 2))
21       if (ww == 0) w = 0 else w = w / ww
22       if (sum(w) > s) lambda = lambda + delta else lambda = lambda -
         delta
23       delta = delta / 2
24     }
25     return(w)
26   }
27
28   comp.a = function(X, y) {
29     n = nrow(X); p = ncol(X); a = array(dim = p)
30     for (j in 1:p) {
31       a[j] = 0
32       for (i in 1:n) for (h in 1:n) a[j] = a[j] + (X[i, j] - X[h, j]) ^
         2 / n
33       for (k in 1:K) {
34         S = 0
35         index = which(y == k)
36         if (length(index) == 0) break
37         for (i in index) for (h in index) S = S + (X[i, j] - X[h, j]) ^
           2
38         a[j] = a[j] - S / length(index)
39       }
40     }
41     return(a)
42   }
```

Example 64 Generating data and executing the function `sparse.k.means`, we observe what variables play crucial roles in clustering.

```
1   p = 10; n = 100; X = matrix(rnorm(p * n), nrow = n, ncol = p)
2   sparse.k.means(X, 5, 1.5)
```

```
1    $w
2    [1] 0.00659343 0.00000000 0.74827158 0.00000000
3    [5] 0.00000000 0.00000000 0.00000000 0.65736124
4    [9] 0.00000000 0.08900768
5    $y
6    [1] 2 2 3 2 2 5 2 5 2 3 1 3 2 4 3 2 1 5 5 4 1 1 2 2
7    [25] 2 3 3 4 2 4 2 2 3 1 1 4 2 3 2 5 2 2 1 1 4 1 4 2
8    [49] 3 1 5 4 5 3 5 4 4 1 3 3 2 3 3 3 3 5 2 1 3 4 3 3
9    [73] 5 4 4 3 5 2 2 5 3 4 3 3 4 1 2 1 3 3 5 5 1 2 4 4
10   [97] 5 4 2 2
```

Because K-means clustering is not convex, we cannot hope to obtain the optimum solution by repeating the functions k.means, a.comp, and w.a. However, we can find the relevant variables for clustering.

7.4   Convex Clustering

Given the data $x_1, \ldots, x_N \in \mathbb{R}^p$, we find $u_1, \ldots, u_N \in \mathbb{R}^p$ that minimize

$$\frac{1}{2} \sum_{i=1}^{N} \|x_i - u_i\|^2 + \gamma \sum_{i<j} w_{i,j} \|u_i - u_j\|_2$$

for $\gamma > 0$, where $w_{i,j}$ is the constant determined from x_i, x_j, such as $\exp(-\|x_i - x_j\|_2^2)$. We refer to this clustering method as convex clustering. If $u_i = u_j$, we regard the samples indexed by i, j as being in the same cluster [7].

To solve via the ADMM, for $U \in \mathbb{R}^{N \times p}$, $V \in \mathbb{R}^{N \times N \times p}$, and $\Lambda \in \mathbb{R}^{N \times N \times p}$, we define the extended Lagrangian as

$$L_v(U, V, \Lambda) := \frac{1}{2} \sum_{i \in V} \|x_i - u_i\|_2^2 + \gamma \sum_{(i,j) \in E} w_{i,j} \|v_{i,j}\| + \sum_{(i,j) \in E} \langle \lambda_{i,j}, v_{i,j} - u_i + u_j \rangle$$

$$+ \frac{v}{2} \sum_{(i,j) \in E} \|v_{i,j} - u_i + u_j\|_2^2 , \tag{7.13}$$

for $u_i \in \mathbb{R}^p$ ($i \in V$), $v_{i,j}, \lambda_{i,j} \in \mathbb{R}^p$ (($i, j) \in E$), where we write (j, k) when the vertices i, j are connected and $j < k$, $V := \{1, \ldots, N\}$ is the set of vertices, and E is a subset of $\{(i, j) \mid i, j \in V, i < j\}$ ($w_{j,i} = w_{i,j}$, $v_{j,i} = v_{i,j}$, $\lambda_{j,i} = \lambda_{i,j}$).

Proposition 24   Suppose that the edge set E is $\{(i, j) \mid i < j, i, j \in V\}$, i.e., all the vertices are connected. If we fix V, Λ and denote

$$y_i := x_i + \sum_{j:(i,j) \in E} (\lambda_{i,j} + v v_{i,j}) - \sum_{j:(j,i) \in E} (\lambda_{j,i} + v v_{j,i}) , \tag{7.14}$$

then (7.13) is minimized when

$$u_i = \frac{y_i + v \sum_{j \in V} x_j}{1 + Nv} . \tag{7.15}$$

For the proof, see the Appendix.

When we fix U, Λ, the minimization of (7.13) w.r.t. V is that of

$$\frac{1}{2} \left\| \frac{1}{v} (\lambda_{i,j} + v v_{i,j}) - (u_i - u_j) \right\|_2^2 + \frac{\gamma}{v} w_{i,j} \|v_{i,j}\|$$

w.r.t. $v_{i,j} \in \mathbb{R}^p$. Therefore, if we write $v \in \mathbb{R}^p$ that minimizes

$$\sigma \Omega(v) + \frac{1}{2}\|u - v\|_2^2$$

by $\text{prox}_{\sigma\Omega}(u)$, (7.13) is minimized w.r.t. V when

$$v_{i,j} = \text{prox}_{\sigma\|\cdot\|}(u_i - u_j - \frac{1}{\nu}\lambda_{i,j}), \tag{7.16}$$

where $\sigma := \dfrac{\gamma w_{i,j}}{\nu}$.

Finally, the update of the Lagrange coefficients is as follows:

$$\lambda_{i,j} = \lambda_{i,j} + \nu(v_{i,j} - u_i + u_j) \tag{7.17}$$

Based on (7.15)–(7.17), we construct the ADMM procedure for the extended Lagrangian (7.13), where we set the weights of the samples such that the distance is more than dd zero.

```
1   ## Computing weights
2   ww = function(x, mu = 1, dd = 0) {
3     n = nrow(x)
4     w = array(dim = c(n, n))
5     for (i in 1:n) for (j in 1:n) w[i, j] = exp(-mu * sum((x[i, ] - x[j,
          ]) ^ 2))
6     if (dd > 0) for (i in 1:n) {
7       dis = NULL
8       for (j in 1:n) dis = c(dis, sqrt(sum((x[i, ] - x[j, ]) ^ 2)))
9       index = which(dis > dd)
10      w[i, index] = 0
11    }
12    return(w)
13  }
14  ## prox (group Lasso) for L2
15  prox = function(x, tau) {
16    if (sum(x ^ 2) == 0) return(x) else return(max(0, 1 - tau / sqrt(sum
          (x ^ 2))) * x)
17  }
18  ## Update u
19  update.u = function(v, lambda) {
20    u = array(dim = c(n, d))
21    z = 0; for (i in 1:n) z = z + x[i, ]
22    y = x
23    for (i in 1:n) {
24      if (i < n) for (j in (i + 1):n) y[i, ] = y[i, ] + lambda[i, j, ] +
          nu * v[i, j, ]
25      if (1 < i) for (j in 1:(i - 1)) y[i, ] = y[i, ] - lambda[j, i, ] -
          nu * v[j, i, ]
26      u[i, ] = (y[i, ] + nu * z) / (n * nu + 1)
27    }
28    return(u)
29  }
```

```
30   ## Update v
31   update.v = function(u, lambda) {
32     v = array(dim = c(n, n, d))
33     for (i in 1:(n - 1)) for (j in (i + 1):n) {
34       v[i, j, ] = prox(u[i, ] - u[j, ] - lambda[i, j, ] / nu, gamma * w[
         i, j] / nu)
35     }
36     return(v)
37   }
38   ## Update lambda
39   update.lambda = function(u, v, lambda) {
40     for (i in 1:(n - 1)) for (j in (i + 1):n) {
41       lambda[i, j, ] = lambda[i, j, ] + nu * (v[i, j, ] - u[i, ] + u[j,
         ])
42     }
43     return(lambda)
44   }
45   ## Repeats the updates of u,v,lambda for the max_iter times
46   convex.cluster = function() {
47     v = array(rnorm(n * n * d), dim = c(n, n, d))
48     lambda = array(rnorm(n * n * d), dim = c(n, n, d))
49     for (iter in 1:max_iter) {
50       u = update.u(v, lambda); v = update.v(u, lambda); lambda = update.
         lambda(u, v, lambda)
51     }
52     return(list(u = u, v = v))
53   }
```

Example 65 Generating data, we execute and cluster. The output is shown in Fig. 7.4. Fix $\nu = 1$, and set the values of γ and tt dd to $(1, 0.5)$, $(10, 0.5)$. Compared to K-means clustering, this method decides whether to join with a pair of

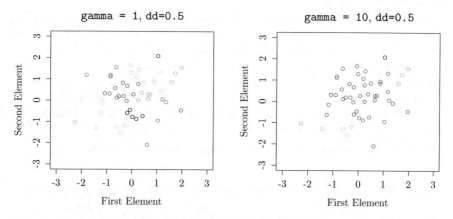

Fig. 7.4 Execution of Example 65. We executed convex clustering with γ and dd being $(1, 0.5)$ and $(10, 0.5)$ for the left and right cases, respectively. Compared with K-means clustering, convex clustering often constructs clusters with distant samples (left)

vertices without realizing the minimum intracluster distribution and the maximum intercluster distribution of the cluster samples.

```
1   ## Data Generation
2   n = 50; d = 2; x = matrix(rnorm(n * d), n, d)
3   ## Convex Clustering
4   w = ww(x, 1, dd = 0.5)
5   gamma=1 # gamma = 10
6   nu = 1; max_iter = 1000; v = convex.cluster()$v
7   ## Adjacency Matrix
8   a = array(0, dim = c(n, n))
9   for (i in 1:(n - 1)) for (j in (i + 1):n) {
10    if (sqrt(sum(v[i, j, ] ^ 2)) < 1 / 10 ^ 4) {a[i, j] = 1; a[j, i] =
        1}
11  }
12  ## Display Figure
13  k = 0
14  y = rep(0, n)
15  for (i in 1:n) {
16    if (y[i] == 0) {
17      k = k + 1
18      y[i] = k
19      if (i < n) for (j in (i + 1):n) if (a[i, j] == 1) y[j] = k
20    }
21  }
22  plot(0, xlim = c(-3, 3), ylim = c(-3, 3), type = "n", main = "gamma =
        10")
23  points(x[, 1], x[, 2], col = y + 1)
```

The clustering will not be uniquely determined even if the cluster size is given, but the cluster size will be biased. Therefore, we use the parameter dd to limit the size of a particular cluster. Therefore, this method guarantees the convexity and efficiency of the calculation, but considering the tuning effort and clustering criteria, it is not always the best method.

Additionally, as for the sparse K-means clustering we considered in Sect. 7.3, for the convex clustering in this section, sparse convex clustering that penalizes extra variables is proposed in (7.13). We consider the following method to be sparse convex clustering (B. Wang et al., 2018) [30]. The purpose is to remove unnecessary variables in the clustering, make its interpretation easier, and save the computation.

Let $x^{(j)}, u^{(j)} \in \mathbb{R}^N$ $(j = 1, \ldots, p)$ be the column vectors of $X, U \in \mathbb{R}^{N \times p}$. We rewrite the first term of (7.13) using the column vectors and add the penalty term $\sum_{j=1}^p r_j \|u^{(j)}\|_2$ for unnecessary variables. For $U \in \mathbb{R}^{N \times p}$, $V \in \mathbb{R}^{N \times N \times p}$, $\Lambda \in \mathbb{R}^{N \times N \times p}$, we define the extended Lagrangian for the ADMM as

$$L_v(U, V, \Lambda) := \frac{1}{2} \sum_{j=1}^p \|x^{(j)} - u^{(j)}\|_2^2 + \gamma_1 \sum_{(i,k) \in E} w_{i,k} \|v_{i,k}\| + \gamma_2 \sum_{j=1}^p r_j \|u^{(j)}\|_2$$

$$+ \sum_{(i,k) \in E} \langle \lambda_{i,k}, v_{i,k} - u_i + u_k \rangle + \frac{v}{2} \sum_{(i,k) \in E} \|v_{i,k} - u_i + u_k\|_2^2,$$

$$(7.18)$$

where r_j is the weight for the penalty and $u_1, \ldots, u_N$ and $u^{(1)}, \ldots, u^{(p)}$ are the row and column vectors of $U \in \mathbb{R}^{N \times p}$, respectively.

The optimization w.r.t. U when V, Λ are fixed is given as follows.

Proposition 25 *Given V, Λ, the $u^{(j)}$ ($j = 1, \ldots, p$) that minimizes (7.18) is given by the $u^{(j)}$ that minimizes*

$$\frac{1}{2} \|G^{-1} y^{(j)} - G u^{(j)}\|_2^2 + \gamma_2 r_j \|u^{(j)}\|_2$$

where $y^{(1)}, \ldots, y^{(p)} \in \mathbb{R}^N$ are the column vectors of $y_1, \ldots, y_N \in \mathbb{R}^p$ in (7.14) and

$$G := \sqrt{1 + Nv}\, I_N - \frac{\sqrt{1 + Nv} - 1}{N} E_N$$

with all the values of $E_N \in \mathbb{R}^{N \times N}$ being ones.

For the proof, see the Appendix.

Note that the inverse matrix of G is given by

$$G^{-1} = \frac{1}{\sqrt{1 + Nv}} \left\{ I_N + \frac{\sqrt{1 + Nv} - 1}{N} E_N \right\}.$$

Proposition 25 suggests solving $G, G^{-1} y^{(j)}, \gamma_2 r_j$ via group Lasso. The original paper centers $u^{(j)}$ in each cycle of the ADMM.

The optimization by V when fixing U, Λ and the optimization by Λ when fixing U, V are the same as in the original convex clustering.

The processing of sparse convex clustering is configured as follows. Other than setting the values of $\gamma_2, r_1, \ldots, r_p$ and defining G, G^{-1}, only the line of tt # # of the function s.update_u is different. In particular, the function gr of the group Lasso defined in Chap. 3 is applied.

```
s.update.u = function(G, G.inv, v, lambda) {
  u = array(dim = c(n, d))
  y = x
  for (i in 1:n) {
    if (i < n) for (j in (i + 1):n) y[i, ] = y[i, ] + lambda[i, j, ] +
      nu * v[i, j, ]
    if (1 < i) for (j in 1:(i - 1)) y[i, ] = y[i, ] - lambda[j, i, ] -
      nu * v[j, i, ]
  }
  for (j in 1:d) u[, j] = gr(G, G.inv %*% y[, j], gamma.2 * r[j])
  for (j in 1:d) u[, j] = u[, j] - mean(u[, j])
  return(u)
}
s.convex.cluster = function() {
  ## Set gamma.2, r[1], ..., r[p]
  G = sqrt(1 + n * nu) * diag(n) - (sqrt(1 + n * nu) - 1) / n * matrix
    (1, n, n)
```

```
15   G.inv = (1 + n * nu) ^ (-0.5) * (diag(n) + (sqrt(1 + n * nu) - 1) /
         n * matrix(1, n, n))
16   v = array(rnorm(n * n * d), dim = c(n, n, d))
17   lambda = array(rnorm(n * n * d), dim = c(n, n, d))
18   for (iter in 1:max_iter) {
19     u = s.update.u(G, G.inv, v, lambda); v = update.v(u, lambda)
20     lambda = update.lambda(u, v, lambda)
21   }
22   return(list(u = u, v = v))
23 }
```

Example 66 Using the function in Example 65 and the functions s.update.u,
+s.convex.cluster + constructed this time, we examine the problem of variable selection by sparse convex clustering. Since we are applying the group Lasso, the N values
contained in each $u^{(j)}$ became 0 at the same time. In Fig. 7.5, the value of γ_2 shows
how the value of a particular sample is reduced. At $\gamma = 1$ and $\gamma = 10$, there was
no change in each $u^{(i)}$ due to *gamma2*. γ_1 has a relationship of u_i, u_j and seems to
have an impact. The execution is carried out according to the code below.

```
1  ## Data Generation
2  n = 50; d = 10; x = matrix(rnorm(n * d), n, d)
3  ## Setting before execution
4  w = ww(x, 1/d, dd = sqrt(d))   ## d is large and adjust it
5  gamma = 10; nu = 1; max_iter = 1000
6  r = rep(1, d)
7  ## Change gamma.2, and execute it, and display the coefficients
8  gamma.2.seq = seq(1, 10, 1)
9  m = length(gamma.2.seq)
10 z = array(dim = c(m, d))
11 h = 0
12 for (gamma.2 in gamma.2.seq) {
13    h = h + 1
14    u = s.convex.cluster()$u
15    print(gamma.2)
16    for (j in 1:d) z[h, j] = u[5, j]
17 }
18 plot(0, xlim = c(1, 10), ylim = c(-2, 2), type = "n",
19      xlab = "gamma.2", ylab = "Coefficients", main = "gamma = 100")
20 for (j in 1:d) lines(gamma.2.seq, z[, j], col = j + 1)
```

Appendix: Proof of Proposition

Proposition 23 *Let $s > 0$ and $a \in \mathbb{R}^p$ with nonnegative elements. Then, the w for*
$\|w\|_2 = 1$ *and* $\|w\|_1 \le s$ *that maximizes* $w^T a$ *can be written as*

$$w = \frac{S_\lambda(a)}{\|S_\lambda(a)\|_2}$$

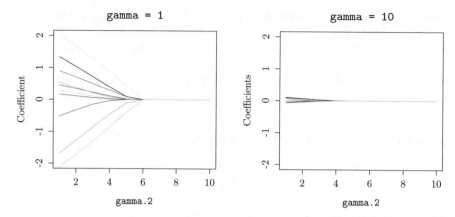

Fig. 7.5 Fixing $\gamma_1 = 1$ (left) and $\gamma_1 = 100$ (right), changing the value of γ_2, we observed the changes in the value of $u_{i,j}$ for a specific sample ($i = 5$). For $\gamma = 10$, the absolute value of the coefficient is reduced compared to that for $\gamma = 1$

for some $\lambda \geq 0$.

Proof: For $w_1, \ldots, w_p$ to minimize the Lagrange

$$L := \sum_{j=1}^{p} -w_j a_j + \frac{\lambda_1}{2}(w^T w - 1) + \lambda_2 \left(\sum_{j=1}^{p} w_j - s \right) - \sum_{j=1}^{p} \mu_j w_j$$

is equivalent to the existence of $\lambda_1, \lambda_2, \mu_1, \ldots, \mu_p \geq 0$ (KKT condition) such that

$$-a_j + \lambda_1 w_j + \lambda_2 - \mu_j = 0 \tag{7.19}$$

$$\mu_j w_j = 0 \tag{7.20}$$

$$w_j \geq 0 \tag{7.21}$$

for $j = 1, \ldots, p$ and

$$w^T w \leq 1 \tag{7.22}$$

$$\sum_{j=1}^{p} w_j \leq s \tag{7.23}$$

$$\lambda_1(w^T w - 1) = 0 \tag{7.24}$$

$$\lambda_2 \left(\sum_{j=1}^{p} w_j - s \right) = 0 . \tag{7.25}$$

Therefore, it is sufficient to show the existence of $\lambda_1, \lambda_2, \mu_1, \ldots, \mu_p$ that satisfy the conditions. First, since $w_j \geq 0$, from (7.19), we have $\lambda_1 w_j = a_j + \mu_j - \lambda_2 \geq 0$. When $a_j - \lambda_2 \geq 0$, if $\mu_j > 0$, then $\lambda_1 w_j > 0$, $\mu_j w_j > 0$, which contradicts (7.20); thus, we have $\mu_j = 0$. When $a_j - \lambda_2 < 0$, from (7.20), we have $\mu_j = \lambda_2 - a_j$ and $w_j = 0$. Therefore, we have $\lambda_1 w = S_{\lambda_2}(a)$. Since $w^T w = 1$, we may choose $\lambda_1 \neq 0$, and the proposition follows. $\square$

Proposition 24 *Suppose that the edge set E is $\{(i, j) \mid i < j, i, j \in V\}$, i.e., all the vertices are connected. If we fix V, Λ and denote*

$$y_i := x_i + \sum_{j:(i,j)\in E} (\lambda_{i,j} + vv_{i,j}) - \sum_{j:(j,i)\in E} (\lambda_{j,i} + vv_{j,i}) \tag{7.14}$$

then (7.13) is minimized when

$$u_i = \frac{y_i + v \sum_{j\in V} x_j}{1 + Nv} . \tag{7.26}$$

Proof The optimization w.r.t. U when V, Λ are fixed is the minimization of

$$f(U) := \frac{1}{2} \sum_{i\in V} \|x_i - u_i\|_2^2 + \frac{v}{2} \sum_{(i,j)\in E} \left\| \frac{1}{v}(\lambda_{i,j} + vv_{i,j}) - u_i + u_j \right\|_2^2 .$$

Note that the term in $f(U)$ that depends on u_i can be written as

$$\frac{v}{2} \left\{ \sum_{j\in V:(i,j)\in E} \left\| \frac{1}{v}(\lambda_{i,j} + vv_{i,j}) - u_i + u_j \right\|_2^2 + \sum_{j\in V:(j,i)\in E} \left\| \frac{1}{v}(\lambda_{i,j} + vv_{i,j}) - u_j + u_i \right\|_2^2 \right\} .$$

If we differentiate $f(U)$ by u_i, we have

$$-(x_i - u_i) - \sum_{j\in V:(i,j)\in E} (\lambda_{i,j} + vv_{i,j} - v(u_i - u_j)) + \sum_{j\in V:(j,i)\in E} (\lambda_{i,j} + vv_{i,j} - v(u_j - u_i)) = 0 ,$$

which is

$$u_i - v \sum_{j\neq i} (u_j - u_i) = x_i + \sum_{j:(i,j)\in E} (\lambda_{i,j} + vv_{i,j}) - \sum_{j:(j,i)\in E} (\lambda_{i,j} + vv_{i,j}) .$$

Thus, we have

$$(Nv + 1)u_i - v \sum_{j\in V} u_j = y_i . \tag{7.27}$$

If we sum both sides of (7.27) over $i \in V$,

$$\sum_{i \in V} u_i = \sum_{i \in V} y_i \qquad (7.28)$$

then from (7.27)(7.28) and $\sum_{j \in V} y_j = \sum_{j \in V} x_j$,

$$u_i = \frac{y_i + v \sum_{j \in V} x_j}{1 + Nv}$$

is the optimum. □

Proposition 25 *Given V, Λ, the $u^{(j)}$ ($j = 1, \ldots, p$) that minimizes (7.18) is given by the $u^{(j)}$ that minimizes*

$$\frac{1}{2} \| G^{-1} y^{(j)} - G u^{(j)} \|_2^2 + \gamma_2 r_j \| u^{(j)} \|_2$$

where $y^{(1)}, \ldots, y^{(p)} \in \mathbb{R}^N$ are the column vectors of $y_1, \ldots, y_N \in \mathbb{R}^p$ in (7.14) and

$$G := \sqrt{1 + Nv} \, I_N - \frac{\sqrt{1 + Nv} - 1}{N} E_N$$

with all the values of $E_N \in \mathbb{R}^{N \times N}$ being ones.

Proof If we differentiate all the terms except the third term in (7.18) by the j-th element of u_i (the i-th element of a_j), then from (7.27), we have

$$(Nv + 1) u_{i,j} - v \sum_{k \in V} u_{k,j} - y_{i,j} , \qquad (7.29)$$

where

$$y_{i,j} := x_{i,j} + \sum_{k:(i,k) \in E} (\lambda_{i,k,j} + vv_{i,k,j}) - \sum_{k:(k,i) \in E} (\lambda_{i,k,j} + vv_{i,k,j}) .$$

If we write (7.29) in the form of a matrix, we have

$$\{(Nv + 1) I_N - v E_N\}[u^{(1)}, \ldots, u^{(p)}] - y ,$$

where y is the matrix such that $y_1, \ldots, y_N \in \mathbb{R}^p$ are the rows, $I_N \in \mathbb{R}^{N \times N}$ is the unit matrix, and $E_N \in \mathbb{R}^{N \times N}$ is the matrix in which all elements are one. Note from $E_N^2 = N E_N$ that $M := (Nv + 1) I_N - v E_N \in \mathbb{R}^{N \times N}$ can be written as the square of the symmetric matrix G. In fact,

$$N \left(\frac{\sqrt{1 + Nv} - 1}{N} \right)^2 - 2 \frac{(\sqrt{1 + vN} - 1)}{N} \cdot \sqrt{1 + Nv} = \frac{1}{N} - \frac{1 + vN}{N} = -v ,$$

and (7.29) can be written as

$$GG[u^{(1)}, \ldots, u^{(p)}] - [y^{(1)}, \ldots, y^{(p)}] .$$

Therefore, if we consider the subderivative of $\gamma_2 \sum_{j=1}^{p} r_j \|u^{(j)}\|_2$, we find that the optimization problem reduces to finding the value of $u^{(j)}$ that minimizes

$$\frac{1}{2}\|G^{-1}y^{(j)} - Gu^{(j)}\|_2^2 + \gamma_2 r_j \|u^{(j)}\|_2 \tag{7.30}$$

for each $j = 1, \ldots, p$. □

Exercises 88–100

In the following, it is assumed that $X \in \mathbb{R}^{N \times p}$ is centered, i.e., $X = (x_{i,j})$, $\sum_{i=1}^{N} x_{i,j} = 0$ ($j = 1, \ldots, p$), and x_i represents a row vector.

88. $V_1 \in \mathbb{R}^p$, which has a length of one and maximizes $\|Xv\|$, is called the first principal component vector. In a vector that is orthogonal from the first principal component vector to the $i - 1$ principal component vector and has a length of one, the $V_i \in \mathbb{R}^p$ that maximizes $\|Xv_i\|$ is called the i principal component vector. We wish to find it for $i = 1, \ldots, m$ ($m \leq p$).

   (a) Show that the first principal component vector is the eigenvector of $\Sigma :=$ $X^T X / N$.
   (b) Show that eigenvectors are orthogonal when the eigenvalues are distinct. Moreover, show that eigenvectors can be orthogonalized. If the eigenvalues are the same, what should we do?
   (c) To obtain the first m principal component vectors, we choose the ones with the largest m eigenvalues and normalize them. Why can we ignore the condition "orthogonal from the first principal component vector to the $i - 1$ principal component vector"?

89. For the principal component vector $V_m := [v_1, \ldots, v_m]$ defined in Exercise 88, show that $(I - V_m V_m^T)^2 = I - V_m V_m^T$. In addition, show that the V_m defined in Exercise 88 and the $V_m \in \mathbb{R}^{N \ timesm}$ that minimizes the reconstruction error $\sum_{i=1}^{N} x_i (I - V_m V_m^T) x_i^T$ coincide.

90. When we find $v \in \mathbb{R}^p$ for $\|v\|_2 = 1$ that maximizes $\|Xv\|_2$, we may restrict the number of nonzero elements in v: maximize $v^T X^T X v - \lambda \|v\|_0\}$ under $\|v\|_0 \leq t$ and $\|v\|_2 = 1$. However, the modification loses the convexity of the problem; thus, we may consider the alternative: the same maximization under the constraint $\|v\|_1 \leq t$ ($t > 0$)

$$\max_{\|v\|_2=1} \{v^T X^T X v - \lambda \|v\|_1\} \quad (\text{cf. 7.4}) \tag{7.31}$$

(SCoTLASS). Nonetheless, it is still not convex. Therefore, we consider the problem

$$\max_{\|u\|_2 = \|v\|_2 = 1} \{u^T X v - \lambda \|v\|_1\} \quad \text{(cf. 7.4)} \tag{7.32}$$

for $u \in \mathbb{R}^N$.

(a) Show that the v obtained in (7.32) optimizes (7.31).

Hint If we differentiate

$$L := -u^T X v + \lambda \|v\|_1 + \frac{\mu}{2}(u^T u - 1) + \frac{\delta}{2}(v^T v - 1) \quad \text{(cf. 7.5)} \tag{7.33}$$

by u, we have $X v - \mu u = 0$, $u = \dfrac{X v}{\|X v\|_2}$. Substitute this into $u^T X v$.

(b) The reason why we transform (7.4) is that we can efficiently obtain the solution of (7.3): Choose an arbitrary $v \in \mathbb{R}^p$ such that $\|v\|_2 = 1$. Repeat to update $u \in \mathbb{R}^N$, $v \in \mathbb{R}^p$:

   i. $u \leftarrow \dfrac{X v}{\|X v\|_2}$

   ii. $v \leftarrow \dfrac{S_\lambda(X^T u)}{\|S_\lambda(X^T u)\|_2}$

until the u, v values converge, where $S_\lambda(z)$ is the function $S_\lambda(\cdot)$ introduced in Chap. 1, which applies each element of $z = [z_1, \cdots, z_p] \in \mathbb{R}^p$. From $\dfrac{\partial L}{\partial u} = \dfrac{\partial L}{\partial v} = 0$, derive $v = \dfrac{S_\lambda(X^T u)}{\|S_\lambda(X^T u)\|_2}$ and $u = \dfrac{X v}{\|X v\|_2}$.

Hint If we take the subderivative of $\|v\|_1$, it is $1, -1, [-1, 1]$ for $v > 0$, $v < 0$, $v = 0$, respectively, which means that

$$\begin{cases} (\text{the } j - \text{th column of } X^T u) - \lambda + \delta v_j = 0, & (\text{the } j - \text{th column of } X^T u) > \lambda \\ (\text{the } j - \text{th column of } X^T u) + \lambda + \delta v_j = 0, & (\text{the } j - \text{th column of } X^T u) < -\lambda \\ (\text{the } j - \text{th column of } X^T u) + \lambda[-1, 1] \ni 0, & -\lambda \le (\text{the } j - \text{th column of } X^T u) \le \lambda \end{cases},$$

for each $j = 1, \ldots, p$, i.e., $\delta v = S_\lambda(X^T u)$.

91. If $\beta \mapsto f(\alpha, \beta)$ and $\alpha \mapsto f(\alpha, \beta)$ are convex w.r.t. $\alpha \in \mathbb{R}^m$ and $\beta \in \mathbb{R}^n$, respectively, we say that the bivariate function $f : \mathbb{R}^m \times \mathbb{R}^n \to \mathbb{R}$ is biconvex.

(a) Show that the bivariate function (7.33) is biconvex w.r.t. u, v.
Hint Because $\|v\|_1$ is convex, differentiate it twice with u_j and v_k, and determine whether the solutions are nonnegative.

(b) In general, biconvexity does not mean convexity. When $N = p = 1$, $X > \sqrt{\mu \delta}$, show that it is not convex.

Hint Because L is a bivariate function of u, v, we have

$$\nabla^2 L = \begin{bmatrix} \dfrac{\partial^2 L}{\partial u^2} & \dfrac{\partial^2 L}{\partial u \partial v} \\ \dfrac{\partial^2 L}{\partial u \partial v} & \dfrac{\partial^2 L}{\partial v^2} \end{bmatrix} = \begin{bmatrix} \mu & -X \\ -X & \delta \end{bmatrix}.$$ If the determinant $\mu\delta - X^2$ is

negative, it contains a negative eigenvalue.

92. In general, if $\Psi : \mathbb{R}^p \times \mathbb{R}^p \to \mathbb{R}$, then $f : \mathbb{R}^p \to \mathbb{R}$ satisfies

$$\begin{cases} f(\beta) \leq \Psi(\beta, \theta), & \theta \in \mathbb{R}^p \\ f(\beta) = \Psi(\beta, \beta) \end{cases}, \qquad \text{(cf. (7.6))}$$

we say that Ψ majorizes f at $\beta \in \mathbb{R}^p$.

(a) Suppose that Ψ majorizes f. Show that if we choose $\beta^0 \in \mathbb{R}^p$ arbitrarily and generate $\beta^1, \beta^2, \ldots$ via

$$\beta^{t+1} = \arg\min_{\beta \in \mathbb{R}^p} \Psi(\beta, \beta^t),$$

we have the following inequality:

$$f(\beta^t) = \Psi(\beta^t, \beta^t) \geq \Psi(\beta^{t+1}, \beta^t) \geq f(\beta^{t+1}).$$

(b) If we define $f(v) := -\|Xv\|_2 + \lambda\|v\|_1$, $\Psi(v, v') := -\dfrac{(Xv)^T(Xv')}{\|Xv'\|_2} + \lambda\|v\|_1$, then show that Ψ majorizes f at v using Schwartz's inequality.

(c) Suppose that we choose v^0 arbitrarily and generate $v^0, v^1, \ldots$ via $v^{t+1} := \arg\max_{\|v\|_2=1} \Psi(v, v^t)$. Express the right-hand side value as X, v, λ. Show also that it coincides with $\dfrac{S_\lambda(X^T u)}{\|S_\lambda(X^T u)\|_2}$.

93. For the procedures below (SCoTLASS), fill in the blank, and execute for the sample.

```
1   soft.th = function(lambda, z) return(sign(z) * pmax(abs(z) -
        lambda, 0))
2   ## Even if z is a vector, soft.th works
3   SCoTLASS = function(lambda, X) {
4       n = nrow(X); p = ncol(X); v = rnorm(p); v = v / norm(v, "2")
5       for (k in 1:200) {
6           u = X %*% v; u = u / norm(u, "2"); v = ## Blank ##
7           v = soft.th(lambda, v); size = norm(v, "2")
8           if (size > 0) v = v / size else break
9       }
10      if (norm(v, "2") == 0) print("All the elements of v are zeros");
            return(v)
11  }
12  ## Sample
```

```
13   n = 100; p = 50; X = matrix(rnorm(n * p), nrow = n); lambda.seq =
        0:10 / 10
14   S = NULL; T = NULL
15   for (lambda in lambda.seq) {
16     v = SCoTLASS(lambda, X)
17     S = c(S, sum(sign(v ^ 2)))
18     T = c(T, norm(X %*% v, "2"))
19   }
20   plot(lambda.seq, S, xlab = "lambda", ylab = "# Nonzero Vectors")
21   plot(lambda.seq, T, xlab = "lambda", ylab = "Variance Sum")
```

94. We formulate a sparse PCA problem different from SCoTLASS (SPCA, sparse principal component analysis):

$$\min_{u,v\in\mathbb{R}^p,\|u\|_2=1}\left\{\frac{1}{N}\sum_{i=1}^N\|x_i - x_i vu^T\|_2^2 + \lambda_1\|v\|_1 + \lambda_2\|v\|_2^2\right\} \quad \text{(cf. 7.8)}$$

$$(7.34)$$

For this method, when we fix u and optimize w.r.t. v, we can apply the elastic net procedure. Show each of the following.

(a) The function (7.34) is biconvex w.r.t. u, v.

Hint When we consider the constraint $\|u\|_2 = 1$, (7.34) can be written as

$$L := \frac{1}{N}\sum_{i=1}^N x_i x_i^T - \frac{2}{N}\sum_{i=1}^N u^T x_i^T x_i v + \frac{1}{N}\sum_{i=1}^N v^T x_i^T x_i v + \lambda_1\|v\|_1 + \lambda_2\|v\|_2 + \mu(u^T u - 1).$$

Show that when differentiating this equation w.r.t. u_j twice, it is nonnegative, and that when differentiating it w.r.t. v_k twice except for $\|v\|_1$, it is still nonnegative.

(b) When we fix v and optimize it w.r.t. u, the optimum solution is given by

$$u = \frac{X^T z}{\|X^T z\|_2},$$

where $z_1 = v^T x_1, \ldots, z_N = v^T x_N$.

Hint Differentiating L w.r.t. u_j, show that the differentiated vector w.r.t. $j = 1, \ldots, p$ satisfies $-\dfrac{2}{N}\sum_{i=1}^N x_i^T x_i v + 2\mu u = 0$.

95. We construct the SPCA process using the R language. The process is repeated to observe how each element of the vector v changes with the number of iterations. Fill in the blanks, execute the process, and display the output as a graph.

```
1  ## Data Generation
2  n = 100; p = 5; x = matrix(rnorm(n * p), ncol = p)
3  ## Compute u,v
4  lambda = 0.001; m = 100
5  g = array(dim = c(m, p))
6  for (j in 1:p) x[, j] = x[, j] - mean(x[, j])
7  for (j in 1:p) x[, j] = x[, j] / sqrt(sum(x[, j] ^ 2))
8  r = rep(0, n)
9  v = rnorm(p)
10 for (h in 1:m) {
11   z = x %*% v
12   u = as.vector(t(x) %*% z)
13   if (sum(u ^ 2) > 0.00001) u = u / sqrt(sum(u ^ 2))
14   for (k in 1:p) {
15     for (i in 1:n) r[i] = sum(u * x[i, ]) - sum(u ^ 2) * sum(x[i,
         -k] * v[-k])
16     S = sum(x[, k] * r) / n
17     v[k] = ## Blank (1) ##
18   }
19   if (sum(v ^ 2) > 0.00001) v = ## Blank (2) ##
20   g[h, ] = v
21 }
22 ## Display Graph
23 g.max = max(g); g.min = min(g)
24 plot(1:m, ylim = c(g.min, g.max), type = "n",
25      xlab = "# Repetitions", ylab = "Each element of v", main = "
         lambda = 0.001")
26 for (j in 1:p) lines(1:m, g[, j], col = j + 1)
```

96. It is not easy to perform sparse PCA on general multiple components instead of the first principal component, and the following is required for SCoTLASS: maximize $u_k^T X v_k$ w.r.t. u_k, v_k under

$$\|v_k\|_2 \leq 1 \,,\ \|v_k\|_1 \leq c \,,\ \|u_k\|_2 \leq 1 \,,\ u_k^T u_j = 0\ (j = 1, \ldots, k-1) \,.$$

Show that the value of u_k when we fix v_k is given by

$$u_k := \frac{P_{k-1}^\perp X v_k}{\|P_{k-1}^\perp X v_k\|_2} \qquad \text{(cf. (7.7))}$$

where $P_{k-1}^\perp := I - \sum_{i=1}^{k-1} u_i u_i^T$.

Hint When $u_1, \ldots, u_{k-1}$ are given, show that $u_j P_{k-1}^\perp X v_k = u_j (I - \sum_{i=1}^{k-1} u_i u_i^T) X v_k = 0$ $(j = 1, \ldots, k-1)$ and that the u_k that maximizes $L = u_k^T X v_k - \mu(u_k^T u_k - 1)$ is $u_k = X v_k / (u_k^T X v_k)$.

97. K-means clustering is a problem of finding the disjoint subsets $C_1, \ldots, C_K$ of $\{1, \ldots, N\}$ that minimize

$$\sum_{k=1}^{K} \sum_{i \in C_k} \|x_i - \bar{x}_k\|_2^2$$

from $x_1, \ldots, x_N \in \mathbb{R}^p$ and the positive integer K, where $\bar{x}_k$ is the arithmetic mean of the samples whose indices are in C_k. Witten–Tibshirani (2010) proposed the formulation that maximizes the weighted sum

$$\sum_{h=1}^{p} w_h \left\{ \frac{1}{N} \sum_{i=1}^{N} \sum_{j=1}^{N} d_{i,j,h}^2 - \sum_{k=1}^{K} \frac{1}{N_k} \sum_{i,j \in C_k} d_{i,j,h}^2 \right\} \tag{7.35}$$

with $w_h \geq 0, h = 1, \ldots, p$, and $d_{i,j,h} := |x_{i,h} - x_{j,h}|$ under $\|w\|_2 \leq 1$, $\|w\|_1 \leq s$, and $w \geq 0$ for $s > 0$, where N_k is the number of samples in C_k.

(a) The following is a general K-means clustering procedure. Fill in the blanks, and execute it.

```
1  k.means = function(X, K, weights = w) {
2    n = nrow(X); p = ncol(X)
3    y = sample(1:K, n, replace = TRUE)
4    center = array(dim = c(K, p))
5    for (h in 1:10) {
6      for (k in 1:K) {
7        if (sum(y[] == k) == 0) center[k, ] = Inf else
8          for (j in 1:p) center[k, j] = ## Blank (1) ##
9      }
10     for (i in 1:n) {
11       S.min = Inf
12       for (k in 1:K) {
13         if (center[k, 1] == Inf) break
14         S = sum((X[i, ] - center[k, ]) ^ 2 * w)
15         if (S < S.min) {S.min = S; ## Blank(2) ##}
16       }
17     }
18   }
19   return(y)
20 }
21 ## Data Generation
22 K = 10; p = 2; n = 1000; X = matrix(rnorm(p * n), nrow = n,
      ncol = p)
23 w = c(1, 1); y = k.means(X, K, w)
24 ## Display Output
25 plot(-3:3, -3:3, xlab = "x", ylab = "y", type = "n")
26 points(X[, 1], X[, 2], col = y + 1)
```

(b) To obtain $w_1, \ldots, w_p$ that maximizes $\sum_{j=1}^{p} w_j a_j$ with

$$a_j(C_1, \ldots, C_K) := \frac{1}{N} \sum_{i=1}^{N} \sum_{i'=1}^{N} (x_{i,j} - x_{i',j})^2 - \sum_{k=1}^{K} \frac{1}{N_k} \sum_{i \in C_k} \sum_{i' \in C_k} (x_{i,j} - x_{i',j})^2 \quad (\text{cf. } (7.11))$$

given $C_1, \ldots, C_K$, we find the value of w that maximizes $\sum_{h=1}^{p} w_h a_h$ under $\|w\|_2 = 1$, $\|w\|_1 \leq s$, and $w \geq 0$ given $a \in \mathbb{R}^p$ with nonnegative elements. We define $\lambda > 0$ as zero and w such that $\|w\|_1 = s$ for the cases where $\|w\|_1 < s$ and $\|w\|_1 = s$. Suppose that $a \in \mathbb{R}^p$ has nonnegative elements.

Based on the fact that

$$w = \frac{\mathcal{S}_\lambda(a)}{\|\mathcal{S}_\lambda(a)\|_2} \qquad\qquad \text{(cf. (7.12))}$$

maximizes $w^T a$ among $w \in \mathbb{R}^p$ with nonnegative elements, we construct the following procedure. Fill in the blanks, and execute it.

```
1   w.a = function(a, s) {
2     a = a/sqrt(sum(a^2))
3     if (sum(a) < s) return(a)
4     p = length(a)
5     lambda = max(a) / 2
6     delta = lambda / 2
7     for (h in 1:10) {
8       for (j in 1:p) w[j] = soft.th(lambda, ## Blank(1) ##)
9       ww = sqrt(sum(w ^ 2))
10      if (ww == 0) w = 0 else w = w / ww
11      if (sum(w) > s) lambda = lambda + delta else lambda = ##
        Blank(2) ##
12      delta = delta / 2
13    }
14    return(w)
15  }
```

98. The following is a sparse K-means procedure based on Exercise 97, to which the functions below have been added. Fill in the blanks, and execute it.

```
1   sparse.k.means = function(X, K, s) {
2     p = ncol(X); w = rep(1, p)
3     for (h in 1:10) {
4       y = k.means(## Blank (1) ##)
5       a = comp.a(## Blank (2) ##)
6       w = w.a(## Blank (3) ##)
7     }
8     return(list(w = w, y = y))
9   }
10  comp.a = function(X, y) {
11    n = nrow(X); p = ncol(X); a = array(dim = p)
12    for (j in 1:p) {
13      a[j] = 0
14      for (i in 1:n) for (h in 1:n) a[j] = a[j] + (X[i, j] - X[h, j
        ]) ^ 2 / n
15      for (k in 1:K) {
16        S = 0
17        index = which(y == k)
18        if (length(index) == 0) break
19        for (i in index) for (h in index) S = S + (X[i, j] - X[h, j
          ]) ^ 2
20        a[j] = a[j] - S / length(index)
21      }
22    }
23    return(a)
24  }
```

```
25   ## Execute the two line below
26   p = 10; n = 100; X = matrix(rnorm(p * n), nrow = n, ncol = p)
27   sparse.k.means(X, 5, 1.5)
```

99. Given data $x_1, \ldots, x_N \in \mathbb{R}^p$ and parameter $\gamma > 0$, find the $u_1, \ldots, u_N \in \mathbb{R}^p$ that minimize

$$\frac{1}{2} \sum_{i=1}^{N} \|x_i - u_i\|^2 + \gamma \sum_{i<j} w_{i,j} \|u_i - u_j\|_2 ,$$

where $w_{i,j}$ are determined from x_i, x_j, such as $\exp(-\|x_i - x_j\|_2^2)$. If $u_i = u_j$, we consider them to be in the same cluster. We define the extended Lagrangian

$$L_\nu(U, V, \Lambda) := \frac{1}{2} \sum_{i \in V} \|x_i - u_i\|_2^2 + \gamma \sum_{(i,j) \in E} w_{i,j} \|v_{i,j}\| + \sum_{(i,j) \in E} \langle \lambda_{i,j}, v_{i,j} - u_i + u_j \rangle$$
$$+ \frac{\nu}{2} \sum_{(i,j) \in E} \|v_{i,j} - u_i + u_j\|_2^2 \qquad \qquad \text{(cf. (7.13))}$$

for $U \in \mathbb{R}^{N \times p}$, $V \in \mathbb{R}^{N \times N \times p}$, and $\Lambda \in \mathbb{R}^{N \times N \times p}$ to obtain the solution of clustering via the ADMM. Fill in the blanks, and execute it. Explain what process is being executed for the optimization w.r.t. U with V, Λ fixed, that w.r.t. V with Λ, U fixed, and that w.r.t. Λ with U, V fixed.

```
1    ww = function(x, mu = 1, dd = 0) {
2      n = nrow(x)
3      w = array(dim = c(n, n))
4      for (i in 1:n) for (j in 1:n) w[i, j] = exp(-mu * sum((x[i, ] -
         x[j, ]) ^ 2))
5      if (dd > 0) for (i in 1:n) {
6        dis = NULL
7        for (j in 1:n) dis = c(dis, sqrt(sum((x[i, ] - x[j, ]) ^ 2)))
8        index = which(dis > dd)
9        w[i, index] = 0
10     }
11     return(w)
12   }
13   prox = function(x, tau) {
14     if (sum(x ^ 2) == 0) return(x) else return(max(0, 1 - tau / sqrt
         (sum(x ^ 2))) * x)
15   }
16   update.u = function(v, lambda) {
17     u = array(dim = c(n, d))
18     z = 0; for (i in 1:n) z = z + x[i, ]
19     y = x
20     for (i in 1:n) {
21       if (i < n) for (j in (i + 1):n) y[i, ] = y[i, ] + lambda[i, j,
           ] + nu * v[i, j, ]
22       if (1 < i) for (j in 1:(i - 1)) y[i, ] = y[i, ] - lambda[j, i,
           ] - nu * v[j, i, ]
23       u[i, ] = (y[i, ] + nu * z) / (n * nu + 1)
24     }
25     return(u)
```

```
26  }
27  update.v = function(u, lambda) {
28    v = array(dim = c(n, n, d))
29    for (i in 1:(n - 1)) for (j in (i + 1):n) {
30      v[i, j, ] = prox(u[i, ] - u[j, ] - lambda[i, j, ] / nu, gamma
        * w[i, j] / nu)
31    }
32    return(v)
33  }
34  update.lambda = function(u, v, lambda) {
35    for (i in 1:(n - 1)) for (j in (i + 1):n) {
36      lambda[i, j, ] = lambda[i, j, ] + nu * (v[i, j, ] - u[i, ] + u
        [j, ])
37    }
38    return(lambda)
39  }
40  ## Repeat the updates of u,v,lambda for max_iter times
41  convex.cluster = function() {
42    v = array(rnorm(n * n * d), dim = c(n, n, d))
43    lambda = array(rnorm(n * n * d), dim = c(n, n, d))
44    for (iter in 1:max_iter) {
45      u = ## Blank (1) ##
46      v = ## Blank (2) ##
47      lambda = ## Blank (3) ##
48    }
49    return(list(u = u, v = v))
50  }
51  ## Data Generation
52  n = 50; d = 2; x = matrix(rnorm(n * d), n, d)
53  ## Convex Clustering
54  w = ww(x, 1, dd = 1); gamma = 10; nu = 1; max_iter = 1000; v =
        convex.cluster()$v
55  ## Adjacency Matrix
56  a = array(0, dim = c(n, n))
57  for (i in 1:(n - 1)) for (j in (i + 1):n) {
58    if (sqrt(sum(v[i, j, ] ^ 2)) < 1 / 10 ^ 4) {a[i, j] = 1; a[j, i]
        = 1}
59  }
60  ## Display
61  k = 0
62  y = rep(0, n)
63  for (i in 1:n) {
64    if (y[i] == 0) {
65      k = k + 1
66      y[i] = k
67      if (i < n) for (j in (i + 1):n) if (a[i, j] == 1) y[j] = k
68    }
69  }
70  plot(0, xlim = c(-3, 3), ylim = c(-3, 3), type = "n", main = "
        gamma = 10")
71  points(x[, 1], x[, 2], col = y + 1)
```

100. Let $x^{(j)}, u^{(j)} \in \mathbb{R}^N$ $(j = 1, \ldots, p)$ be the column vectors of $X, U \in \mathbb{R}^{N \times p}$. We define the extended Lagrangian as

$$L_v(U, V, \Lambda) := \frac{1}{2} \sum_{j=1}^{p} \|x^{(j)} - u^{(j)}\|_2^2 + \gamma_1 \sum_{(i,k)\in E} w_{i,k} \|v_{i,k}\| + \gamma_2 \sum_{j=1}^{p} r_j \|u^{(j)}\|_2$$

$$+ \sum_{(i,k)\in E} \langle \lambda_{i,k}, v_{i,k} - u_i + u_k \rangle + \frac{v}{2} \sum_{(i,k)\in E} \|v_{i,k} - u_i + u_k\|_2^2 \quad \text{(cf. (7.18))}$$

for convex optimization w.r.t. $U \in \mathbb{R}^{N\times p}$, $V \in \mathbb{R}^{N\times N\times p}$, $\Lambda \in \mathbb{R}^{N\times N\times p}$ (sparse convex clustering), where r_j is the weight for the penalty of the j-th variable. The following is the code. Unlike ordinary convex clustering, the parameters $\gamma_2, r_1, \ldots, r_p$ have been added, and the functions s.update.u and s.convex.cluster have been modified. Explain why these differences are required between convex clustering and sparse convex clustering.

```
1   s.update.u = function(v, lambda) {
2     u = array(dim = c(n, d))
3     y = x
4     for (i in 1:n) {
5       if (i < n) for (j in (i + 1):n) y[i, ] = y[i, ] + lambda[i, j,
            ] + nu * v[i, j, ]
6       if (1 < i) for (j in 1:(i - 1)) y[i, ] = y[i, ] - lambda[j, i,
            ] - nu * v[j, i, ]
7     }
8     for (j in 1:d) u[, j] = gr(G, G.inv %*% y[, j], gamma.2 * r[j])
9     for (j in 1:d) u[, j] = u[, j] - mean(u[, j])
10    return(u)
11  }
12  s.convex.cluster = function() {
13    ## gamma.2, r[1], ..., r[p]
14    G = sqrt(1 + n * nu) * diag(n) - (sqrt(1 + n * nu) - 1) / n %*%
          matrix(1, n, n)
15    G.inv = (1 + n * nu) ^ (-0.5) %*%
16      (diag(n) + (sqrt(1 + n * nu) - 1) / n * matrix(1, n, n))
17    v = array(rnorm(n * n * d), dim = c(n, n, d))
18    lambda = array(rnorm(n * n * d), dim = c(n, n, d))
19    for (iter in 1:max_iter) {
20      u = s.update.u(v, lambda); v = update.v(u, lambda)
21      lambda = update.lambda(u, v, lambda)
22    }
23    return(list(u = u, v = v))
24  }
```

Bibliography

1. Alizadeh, A., Eisen, M., Davis, R.E., Ma, C., Lossos, I., Rosenwal, A., Boldrick, J., Sabet, H., Tran, T., Yu, X., Pwellm, J., Marti, G., Moore, T., Hudsom, J., Lu, L., Lewis, D., Tibshirani, R., Sherlock, G., Chan, W., Greiner, T., Weisenburger, D., Armitage, K., Levy, R., Wilson, W., Greve, M., Byrd, J., Botstein, D., Brown, P., Staudt, L.: Identification of molecularly and clinically distinct subtypes of diffuse large B cell lymphoma by gene expression profiling. Nature **403**, 503–511 (2000)
2. Arnold, T., Tibshirani, R.: genlasso: path algorithm for generalized lasso problems. R package version 1.5
3. Beck, A., Teboulle, M.: A fast iterative shrinkage-thresholding algorithm for linear inverse problems. SIAM J. Imaging Sci. **2**, 183–202 (2009)
4. Bertsekas, D.: Convex Analysis and Optimization. Athena Scientific, Nashua (2003)
5. Boyd, S., Parikh, N., Chu, E., Peleato, B., Eckstein, J.: Distributed optimization and statistical learning via the alternating direction method of multipliers. Found. Trends Mach. Learn. **3**(1), 1–124 (2011)
6. Boyd, S., Vandenberghe, L.: Convex Optimization. Cambridge University Press, Cambridge (2004)
7. Chi, E.C., Lange, K.: Splitting methods for convex clustering. J. Comput. Graph. Stat. (online access) (2014)
8. Danaher, P., Witten, D.: The joint graphical lasso for inverse covariance estimation across multiple classes. J. R. Stat. Soc. Ser. B **76**(2), 373–397 (2014)
9. Efron, B., Hastie, T., Johnstone, I., Tibshirani, B.: Least angle regression. Ann. Stat. **32**(2), 407–499 (2004)
10. Friedman, J., Hastie, T., Hoefling, H., Tibshirani, R.: Pathwise coordinate optimization. Ann. Appl. Stat. **1**(2), 302–332 (2007)
11. Friedman, J., Hastie, T., Simon, N., Tibshirani, R.: glmnet: lasso and elastic-net regularized generalized linear models. R package version 4.0 (2015)
12. Friedman, J., Hastie, T., Tibshirani, R.: Sparse inverse covariance estimation with the graphical lasso. Biostatistics **9**, 432–441 (2008)
13. Jacob, L., Obozinski, G., Vert, J.-P.: Group lasso with overlap and graph lasso. In: Proceeding of the 26th International Conference on Machine Learning, Montreal, Canada (2009)
14. Johnson, N.: A dynamic programming algorithm for the fused lasso and '0-segmentation. J. Comput. Graph. Stat. **22**(2), 246–260 (2013)

15. Jolliffe, I.T., Trendafilov, N.T., Uddin, M.: A modified principal component technique based on the lasso. J. Comput. Graph. Stat. **12**, 531–547 (2003)
16. Suzuki, J.: Statistical Learning with Math and R. Springer, Berlin (2020)
17. Kawano, S., Matsui,H., Hirose, K.: Statistical Modeling for Sparse Estimation. Kyoritsu-Shuppan, Tokyo (2018) (in Japanese)
18. Lauritzen, S.L.: Graphical Models. Oxford University Press, Oxford (1996)
19. Mazumder, R., Hastie, T., Tibshirani, R.: Spectral regularization algorithms for learning large incomplete matrices. J. Mach. Learn. Res. **11**, 2287–2322 (2010)
20. Meinshausen, N., Bühlmann, P.: High-dimensional graphs and variable selection with the lasso. Ann. Stat. **34**, 14361462 (2006)
21. Mota, J., Xavier, J., Aguiar, P., Püschel, M.: A proof of convergence for the alternating direction method of multipliers applied to polyhedral-constrained functions. Optimization and Control, Mathematics arXiv (2011)
22. Nesterov, Y.: Gradient methods for minimizing composite objective function. Technical Report 76, Center for Operations Research and Econometrics (CORE), Catholic University of Louvain (UCL) (2007)
23. Ravikumar, P., Liu, H., Lafferty, J., Wasserman, L.: Sparse additive models. J. R. Stat. Soc. Ser. B **71**(5), 1009–1030 (2009)
24. Ravikumar, P., Wainwright, M.J., Raskutti, G., Yu, B.: High-dimensional covariance estimation by minimizing 1-penalized logdeterminant divergence. Electron. J. Stat. **5**, 935–980 (2011)
25. Simon, N., Friedman, J., Hastie, T., Tibshirani, R.: Regularization paths for Cox's proportional hazards model via coordinate descent. J. Stat. Softw. **39**(5), 1–13 (2011)
26. Simon, N., Friedman, J., Hastie, T., Tibshirani, R.: A sparse-group lasso. J. Comput. Graph. Stat. **22**(2), 231–245 (2013)
27. Simon, N., Friedman, J., Hastie, T.: A blockwise descent algorithm for group-penalized multiresponse and multinomial regression. Computation, Mathematics arXiv (2013)
28. Tibshirani, R.: Regression shrinkage and selection via the lasso. J. R. Stat. Soc. Ser. B **58**, 267–288 (1996)
29. Tibshirani, R., Taylor, J.: The solution path of the generalized lasso. Ann. Stat. **39**(3), 1335–1371 (2011)
30. Wang, B., Zhang, Y., Sun, W., Fang, Y.: Sparse convex clustering. J. Comput. Graph. Stat. **27**(2), 393–403 (2018)
31. Witten, D., Tibshirani, R.: A framework for feature selection in clustering. J. Am. Stat. Assoc. **105**(490), 713–726 (2010)
32. Zou, H., Hastie, T., Tibshirani, R.: Sparse principal component analysis. J. Comput. Graph. Stat. **15**(2), 265–286 (2006)

Printed in the United States
by Baker & Taylor Publisher Services